INTERNATIONAL CENTRE FOR MECHANICAL SCIENCES

COURSES AND LECTURES - No 29

IMRE CSISZÁR

MATHEMATICAL INSTITUTE
HUNGARIAN ACADEMY OF SCIENCES, BUDAPEST

CHANNEL CODING THEORY

COURSE HELD AT THE DEPARTMENT
FOR AUTOMATION AND INFORMATION
JULY 1970

UDINE 1970

SPRINGER-VERLAG WIEN GMBH

Originally published by Springer-Verlag Wien New York in 1970

ISBN 978-3-211-81089-7 ISBN 978-3-7091-2724-7 (eBook)

DOI 10.1007/978-3-7091-2724-7

PREFACE

Mathematical information theory has been developed in order to investigate the possibilities of reliable communication over channels subject to noise. Although at present the scope of information theory is considerably wider than that, the study of the above problem still keeps to be its central part.

If the characteristics of the channel are given, the only way of increasing the efficiency and reliability of communication is to use proper encoding and decoding methods. As a rule, the first step is to represent the output of the information source in some standard form (source coding); then, before entering the channel, the messages are encoded in order to be protected against noise (channel coding).At the output of the channel, the corresponding decoding operations are performed.

As regards to channel coding, the knowledge of the optimum performance of such techniques and of how to implement encoding and decoding such as to perform not much worse than the theoretical optimum are both important. In the sequel, we shall concentrate on the first problem, outlining the main existence theorems of channel coding theory for discrete one-way channels.

These notes represent the material of the author's lectures at the CISM's Summer course in Udine, 1970.

The author is indebted to Prof. L. Sobrero, Secretary General of CISM, for having invited him to give these lectures and also to Prof. G. Longo whose enthusiastic work in organizing this information theory course was a main factor of its success.

Udine, July 1970

Imre Csiszár

Preliminaries

In this section we summarize some basic definitions and relations which will be used freely in the sequel : the simple proofs will be sketched only.

The term "random variable" will be abbreviated as RV ; for the sake of simplicity, attention will be restricted to the case of discrete RV's, i.e., to RV's with values in a finite or countably infinite set.

ξ, η, ζ will denote RV's with values in the (finite or countably infinite) sets X, Y, Z.

All random variables considered at the same time will be assumed to be defined on the same probability space. Recall that a probability space is a triplet $(\Omega, \mathfrak{F}, P)$ where Ω is a set (the set of all conceivable outcomes of an experiment), $\mathfrak{F}$ is a σ-algebra of subsets of Ω (the class of observable events) and P is a measure (non-negative countably additive set function) defined on $\mathfrak{F}$ such that $P(\Omega)=1$. RV's ξ, η etc. are functions $\xi(\omega)$, $\eta(\omega)$ etc. $(\omega \in \Omega)$. The probability $P\{\xi = x\}$ is the measure of the set of those ω's for which $\xi(\omega) = x$; similarly, $P\{\xi = x, \eta = y\}$ is the measure of the set of those ω's for which $\xi(\omega) = x$ and $\eta(\omega) = y$.

The conditional probability $P\{\xi=x \mid \eta=y\}$ is defined as $\frac{P\{\xi=x,\eta=y\}}{P\{\eta=y\}}$ (if $P\{\eta=y\}=0$, $P\{\xi=x\mid\eta=y\}$ is undefined).

__Definition 1__. The RV's defined by

$$(1) \quad \iota_\xi = -\log_2 P\{\xi=x\} \qquad \text{if} \quad \xi=x$$

$$(2) \quad \iota_{\xi\wedge\eta} = \log_2 \frac{P\{\xi=x,\eta=y\}}{P\{\xi=x\}P\{\eta=y\}} \qquad \text{if} \quad \xi=x,\ \eta=y$$

are called the __entropy density__ of ξ and the __information density__ of ξ and η, respectively.

$$(3) \quad \iota_{\xi|\eta} = -\log_2 P\{\xi=x\mid\eta=y\} \qquad \text{if} \quad \xi=x,\ \eta=y$$

$$(4) \quad \iota_{\xi|\eta,\zeta} = -\log_2 P\{\xi=x\mid\eta=y,\zeta=z\} \quad \text{if} \quad \xi=x,\ \eta=y,\ \zeta=z$$

are __conditional entropy densities__ and

$$(5) \quad \iota_{\xi\wedge\eta|\zeta} = \log_2 \frac{P\{\xi=x,\eta=y\mid\zeta=z\}}{P\{\xi=x\mid\zeta=z\}P\{\eta=y\mid\zeta=z\}} \quad \text{if} \quad \xi=x,\eta=y,\zeta=z$$

__is a conditional information density.__

__Remark__. Entropy density is often called "self-information" and information density "mutual information". In our terminology, the latter term will mean the expectation of $\iota_{\xi\wedge\eta}$.

Definition 2. The quantities

$$H(\xi) \overset{\text{def}}{=} E\iota_{\xi} = -\sum_{x\in X} P\{\xi=x\} \log_2 P\{\xi=x\} \tag{6}$$

$$I(\xi\wedge\eta) \overset{\text{def}}{=} E\iota_{\xi\wedge\eta} = \sum_{x\in X, y\in Y} P\{\xi=x,\eta=y\} \log_2 \frac{P\{\xi=x,\eta=y\}}{P\{\xi=x\}P\{\eta=y\}} \tag{7}$$

are called the entropy of ξ and the mutual information of ξ and η, respectively.

The quantities

$$H(\xi|\eta) \overset{\text{def}}{=} E\iota_{\xi|\eta} = -\sum_{x\in X, y\in Y} P\{\xi=x,\eta=y\} \log_2 P\{\xi=x|\eta=y\} \tag{8}$$

$$H(\xi|\eta,\zeta) \overset{\text{def}}{=} E\iota_{\xi|\eta,\zeta} = -\sum_{x\in X, y\in Y, z\in Z} P\{\xi=x,\eta=y,\zeta=z\} \log_2 P\{\xi=x|\eta=y,\zeta=z\} \tag{9}$$

are called conditional entropies and

$$I(\xi\wedge\eta|\zeta) \overset{\text{def}}{=} E\iota_{\xi\wedge\eta|\zeta} = \sum_{x\in X, y\in Y, z\in Z} P\{\xi=x,\eta=y,\zeta=z\} \log_2 \frac{P\{\xi=x,\eta=y|\zeta=z\}}{P\{\xi=x|\zeta=z\}P\{\eta=y|\zeta=z\}} \tag{10}$$

is called conditional mutual information.

Here terms like $0\log_2 0$ or $0\log_2 \frac{0}{0}$ are meant to be 0.

The quantities (6)-(10) are always non-negative (for (7) and (10) this requires proof ; see (17), (18)) but they may be infinite. The latter contingency should be kept in mind ; in particular, identities like $I(\xi\wedge\eta) = H(\xi) - H(\xi|\eta)$ (cf. (21)) are valid only under the condition that they do not contain the undefined expression $+\infty - \infty$.

$H(\xi)$ is interpreted as the measure of the average amount of information contained in specifying a particular value of ξ; $I(\xi \wedge \eta)$ is a measure of the average amount of information obtained with respect to the value of η when specifying a particular value of ξ. Conditional entropy and conditional mutual information are interpreted similarly. Logarithms to the basis 2 (rather than natural logarithms) are used to ensure that the amount of information provided by a binary digit (more exactly, by a random variable taking on the values 0 and 1 with probabilities 1/2) be unity. This unit of the amount of information is called <u>bit</u>.

The interpretation of the quantities (6)-(10) as measures of the amount of information is not merely a matter of convention ; rather, it is convincingly suggested by a number of theorems of information theory as well as by the great efficency of heuristic reasonings based on this interpretation. There is much less evidence for a similar interpretation of the entropy and information densities. Thus we do not insist on attaching any intuitive meaning to the latters ; they will be used simply as convenient mathematical tools.

A <u>probability distribution</u>, to be abbreviated as PD, on the set X is a non-negative valued function $p(x)$ on X with $\sum_{x \in X} p(x) = 1$; PD's will be denoted by script letters, e.g. $\mathcal{P} = \{p(x),\ x \in X\}$.

Definition 3. The I-divergence of two PD's $\mathcal{P} = \{p(x),\ x \in X\}$ and $\mathcal{Q} = \{q(x), x \in X\}$ is defined as

$$I(\mathcal{P} \| \mathcal{Q}) = \sum_{x \in X} p(x) \log_2 \frac{p(x)}{q(x)} \quad . \tag{11}$$

Here terms of the form $a \log_2 \frac{a}{0}$ with $a > 0$ are meant to be $+\infty$.

Lemma 1. Using the notations $p(A) = \sum_{x \in A} p(x)$, $q(A) = \sum_{x \in A} q(x)$ we have for an arbitrary subset A of X

$$\sum_{x \in A} p(x) \log_2 \frac{p(x)}{q(x)} \geq p(A) \log_2 \frac{p(A)}{q(A)} \ ; \tag{12}$$

if $q(A) > 0$ the equality holds iff(*) $p(x) = \frac{p(A)}{q(A)} q(x)$ for every $x \in A$. In particular, setting $A = X$:

$$I(\mathcal{P} \| \mathcal{Q}) \geq 0 \ , \qquad \text{equality iff } \mathcal{P} = \mathcal{Q} . \tag{13}$$

Proof. The concavity of the function $f(t) = \ln t$ implies $\ln t \leq t - 1$, with equality iff $t = 1$. Setting now $t = \frac{q(x)}{p(x)} \frac{p(A)}{q(A)}$ one gets $\ln \frac{q(x)}{p(x)} \leq \ln \frac{q(A)}{p(A)} + \frac{q(x)}{p(x)} \frac{p(A)}{q(A)} - 1$ whenever $p(x)\,q(x) > 0$, with equality iff $\frac{q(x)}{p(x)} = \frac{q(A)}{p(A)}$. Multiplying by $p(x)$ and summing for every $x \in A$ with $p(x) > 0$ (one may obviously assume that then $q(x) > 0$ too) (12)follows, including the condition for equality. The choice of the basis of the logarithms being clearly immaterial. The I-divergence $I(\mathcal{P} \| \mathcal{Q})$ is a measure of how different the PD $\mathcal{P}$ is from the PD $\mathcal{Q}$ (however note, that in general $I(\mathcal{P} \| \mathcal{Q}) = I(\mathcal{Q} \| \mathcal{P})$). If $\mathcal{P}$ and $\mathcal{Q}$ are two

(*)Iff is an abbreviation for "if and only if".

hypothetical PD's on X then $I(\mathcal{P}\|\mathcal{Q})$ may be interpreted as the average amount of information in favour of $\mathcal{P}$ and against $\mathcal{Q}$, obtained from observing a randomly chosen element of X, provided that the PD $\mathcal{P}$ is the true one.

The <u>distribution of a RV</u> ξ is the PD $\mathcal{P}_\xi$ defined by

(14) $$\mathcal{P}_\xi = \{p_\xi(x),\ x \in X\},\quad p_\xi(x) = P\{\xi = x\}.$$

The <u>joint distribution</u> $\mathcal{P}_{\xi\eta}$ of the RV's ξ and η is defined as the distribution of the RV (ξ,η) taking values in $X \times Y$ i.e. $\mathcal{P}_{\xi\eta} = \{p_{\xi\eta}(x,y),\ x\in X, y\in Y\}$, $p_{\xi\eta}(x,y) = P\{\xi = x,\ \eta = y\}$.

From (7) and (11) it follows

(15) $$I(\xi\wedge\eta) = I(\eta\wedge\xi) = I(\mathcal{P}_{\xi\eta}\,\|\,\mathcal{P}_\xi\times\mathcal{P}_\eta)$$

where $\mathcal{P}_\xi\times\mathcal{P}_\eta = \{p_\xi(x)\,p_\eta(y),\ x\in X, y\in Y\}$ and also

(16) $$I(\xi\wedge\eta) = \sum_{x\in X} p_\xi(x)\, I(\mathcal{P}_{\eta|\xi=x}\,\|\,\mathcal{P}_\eta)$$

where $\mathcal{P}_{\eta|\xi=x} = \{p_x(y),\ y\in Y\}$, $p_x(y) = P\{\eta = y \quad \xi = x\}$.
(15) and (13) yield

(17) $I(\xi\wedge\eta) \geq 0$, equality iff ξ and η are independent.

By a comparison of (7) and (10), this implies

(18) $I(\xi\wedge\eta\,|\,\zeta) \geq 0$, equality iff ξ and η are condition

ally independent for ζ given.

Let us agree to write $\iota_{\xi,\eta}$ for $\iota_{(\xi,\eta)}$ (entropy density of the RV (ξ,η)), $\iota_{\xi,\eta\wedge\zeta}$ for $\iota_{(\xi,\eta)\wedge\zeta}$ (information density of the RV's (ξ,η) and ζ) etc. ; omitting the brackets will cause no ambiguities.

Theorem 1. (Basic identities)

$$\iota_{\xi,\eta} = \iota_{\xi|\eta} + \iota_{\eta} \qquad H(\xi,\eta) = H(\xi|\eta) + H(\eta) \tag{19}$$

$$\iota_{\xi,\eta|\zeta} = \iota_{\xi|\eta,\zeta} + \iota_{\eta|\zeta} \qquad H(\xi,\eta|\zeta) = H(\xi|\eta,\zeta) + H(\eta|\zeta) \tag{20}$$

$$\iota_{\xi} = \iota_{\xi|\eta} + \iota_{\xi\wedge\eta} \qquad H(\xi) = H(\xi|\eta) + I(\xi\wedge\eta) \tag{21}$$

$$\iota_{\xi|\zeta} = \iota_{\xi|\eta,\zeta} + \iota_{\xi\wedge\eta|\zeta} \qquad H(\xi|\zeta) = H(\xi|\eta,\zeta) + I(\xi\wedge\eta|\zeta) \tag{22}$$

$$\iota_{\xi_1,\xi_2\wedge\eta} = \iota_{\xi_1\wedge\eta} + \iota_{\xi_2\wedge\eta|\xi_1} ; \quad I(\xi_1,\xi_2\wedge\eta) = I(\xi_1\wedge\eta) + I(\xi_2\wedge\eta|\xi_1) \tag{23}$$

$$\iota_{\xi_1,\xi_2\wedge\eta|\zeta} = \iota_{\xi_1\wedge\eta|\zeta} + \iota_{\xi_2\wedge\eta|\xi_1,\zeta} ; \quad I(\xi_1,\xi_2\wedge\eta|\zeta) = I(\xi_1\wedge\eta|\zeta) + I(\xi_2\wedge\eta\ \xi_1,\zeta) \tag{24}$$

Proof. Immediate from definitions 1 and 2.

Theorem 2. (Basic inequalities)

The information quantities (6)-(10) are non-negative ;

$$H(\xi,\eta) \geq H(\xi) , \quad H(\xi,\eta|\zeta) \geq H(\xi|\zeta) \tag{25}$$

$$H(\xi|\eta,\zeta) \leq H(\xi|\eta) \leq H(\xi) \tag{26}$$

$$I(\xi_1,\xi_2\wedge\eta)\geqq I(\xi_1\wedge\eta);\ I(\xi_1,\xi_2\wedge\eta|\zeta)\geqq$$

(27) $$\geqq I(\xi_1\wedge\eta|\zeta)$$

(28) $$I(\xi\wedge\eta)\leqq H(\xi),\ I(\xi\wedge\eta|\zeta)\leqq H(\xi|\zeta).$$

If ξ has at most r possible values then

(29) $$H(\xi)\leqq \log_2 r.$$

If ξ has at most $r(y)$ possible values when $\eta=y$ then

(30) $$H(\xi|\eta)\leqq E\log_2 r(\eta).$$

Proof. (25)-(28) are direct consequences of (19)-(24). (29) follows from (13) setting $\mathcal{P}=\mathcal{P}_\xi$, $\mathcal{Q}=\left\{\frac{1}{r},\ldots,\frac{1}{r}\right\}$; on comparison of (6) and (8), (29) implies (30).

Remark. $I(\xi\wedge\eta|\zeta)\leqq I(\xi\wedge\eta)$ is not valid; in general. E. g., if ξ and η are independent but not conditionally independent for a given ζ, then

$$I(\xi\wedge\eta)=0<I(\xi\wedge\eta|\zeta).$$

Theorem 3. (Substitutions in the information quantities).

For arbitrary functions $f(x)$, $f(y)$ or $f(x,y)$ defined on X, Y or $X\times Y$, respectively, the following inequalities hold

(31) $$H(f(\xi))\leqq H(\xi);\quad I(f(\xi)\wedge\eta)\leqq I(\xi\wedge\eta)$$

$$H(\xi\,|\,f(\eta)) \geq H(\xi\,|\,\eta) \tag{32}$$

$$H(f(\xi,\eta)\,|\,\eta) \leq H(\xi\,|\,\eta). \tag{33}$$

If f is one-to-one, or $f(x,y)$ as a function of x is one-to-one for every fixed $y\in Y$, respectively, the equality signs are valid. In the second half of (31) and in (32) the equality holds also if ξ and η are conditionally independent for given $f(\xi)$ or $f(\eta)$, respectively.

Proof. In the one-to-one case, the validity of (31)-(33) with the equality sign is obvious from definition 2. In the general case, apply this observation for $\tilde{f}$ instead of f where $\tilde{f}(x)=(x,f(x))$, $\tilde{f}(y)=(y,f(y))$ or $\tilde{f}(x,y)=(x,f(x,y))$, respectively ; then theorem 2 gives rise to the desired inequalities. The last statements follow from (18) and the identities :

$$I(\xi\wedge\eta) = I(\xi,f(\xi)\wedge\eta) = I(f(\xi)\wedge\eta) + I(\xi\wedge\eta\ f(\xi))$$

$$H(\xi) = H(\xi,f(\xi)) \geq H(f(\xi))$$

$$H(\xi\,|\,\eta) = H(\xi\,|\,\eta,f(\eta)) \leq H(\xi\,|\,f(\eta))$$

$$H(\xi\,|\,\eta) = H(\xi,f(\xi,\eta)\,|\,\eta) \geq H(f(\xi,\eta)\,|\,\eta)$$

respectively.

Theorem 4. (Convexity properties).

Consider the entropy and the mutual information as a function of the distribution of ξ, in the latter case keeping the conditional distributions $\mathcal{P}_{\eta|\xi=x} = \{p_x(y), y \in Y\}$ fixed :

(34) $$H(\mathcal{P}) = -\sum_{x \in X} p(x) \log_2 p(x)$$

(35) $$I(\mathcal{P}) = \sum_{x \in X, y \in Y} p(x) p_x(y) \log_2 \frac{p_x(y)}{q_{\mathcal{P}}(y)} ; \quad q_p(y) = \sum_{x \in X} p(x) p_x(y).$$

Then $H(\mathcal{P})$ and $I(\mathcal{P})$ are concave functions of the PD $\mathcal{P} = \{p(x), x \in X\}$ i.e., if $\mathcal{P}_1 = \{p_1(x) = x \in X\}$, $\mathcal{P}_2 = \{p_2(x), x \in X\}$ and $\mathcal{P} = a\mathcal{P}_1 + (1-a)\mathcal{P}_2 = \{a p_1(x) + (1-a) p_2(x), x \in X\}$ where $0 < a < 1$ is arbitrary, then

(36) $$H(\mathcal{P}) \geq a H(\mathcal{P}_1) + (1-a) H(\mathcal{P}_2), \quad I(\mathcal{P}) \geq a I(\mathcal{P}_1) + (1-a) I(\mathcal{P}_2).$$

Proof. The function $f(t) = -t \log_2 t$ is concave hence so is $H(\mathcal{P})$ as well. Since the PD $\mathcal{Q}_{\mathcal{P}} = \{q_p(y), y \in Y\}$ depends linearly on the PD $\mathcal{P}$, the concavity of $f(t) = -t \log_2 t$ also implies that

$$\sum_{x \in X} p(x) p_x(y) \log_2 \frac{p_x(y)}{q_p(y)} =$$

$$= -q_p(y) \log_2 q_{\mathcal{P}}(y) + \sum_{x \in X} p(x) p_x(y) \log_2 p_x(y)$$

is a concave function of $\mathcal{P}$, for every fixed $y \in Y$. Summation for all $y \in Y$ shows that $I(\mathcal{P})$ is concave, too.

Theorem 5. (Useful estimates with the I-divergence).

Let $\mathcal{P} = \{p(x), x \in X\}$ and $\mathcal{Q} = \{q(x), x \in X\}$ be two PD's on X. Then

$$\sum_{x \in X} |p(x) - q(x)| \leq \sqrt{\frac{2}{\log_2 e} I(\mathcal{P}\|\mathcal{Q})} \tag{37}$$

$$\sum_{x \in X} p(x) \left| \log_2 \frac{p(x)}{q(x)} \right| \leq I(\mathcal{P}\|\mathcal{Q}) + \min\left(\frac{2\log_2 e}{e}, \sqrt{2\log_2 e \cdot I(\mathcal{P}\|\mathcal{Q})}\right) \tag{38}$$

<u>Proof</u>. Let $A = \{x : p(x) \leq q(x)\}$, $B = \{x : p(x) > q(x)\}$; put $p(A) = p$, $q(A) = q$. Then $p \leq q$, $p(B) = 1-p$, $q(B) = 1-q$,

$$\sum_{x \in X} |p(x) - q(x)| = 2(q-p), \tag{39}$$

while from (11) and (12) it follows

$$I(\mathcal{P}\|\mathcal{Q}) \geq p \log_2 \frac{p}{q} + (1-p) \log_2 \frac{1-p}{1-q} . \tag{40}$$

A simple calculation shows that

$$p \log_2 \frac{p}{q} + (1-p)\log_2 \frac{1-p}{1-q} - 2\log_2 e \cdot (p-q)^2 \geq 0 \qquad (0 \leq p \leq q \leq 1) \tag{41}$$

(For $p = q$ the equality holds and the derivative of the left hand side of (41) with respect to p is ≤ 0 if $0 < p \leq q < 1$).

The relations (39), (40), (41) prove (37).

From (11) and (12) it also follows

$$\sum_{x \in X} p(x) \left| \log_2 \frac{p(x)}{q(x)} \right| = I(\mathcal{P}\|\mathcal{Q}) - 2 \sum_{x \in A} p(x) \log_2 \frac{p(x)}{q(x)} \leq$$

$$\leq I(\mathcal{P}\|\mathcal{Q}) - 2p \log_2 \frac{p}{q} = I(\mathcal{P}\|\mathcal{Q}) + 2p \log_2 \frac{q}{p} . \tag{42}$$

Here

$$2p\log_2\frac{q}{p} = 2\log_2 e\cdot p\ln\frac{q}{p} \leqq 2\log_2 e\cdot p\ln\frac{1}{p} \leqq \frac{2\log_2 e}{e}$$

(since $f(t) = t\ln\frac{1}{t}$ takes on its maximum for $t=\frac{1}{e}$); furthermore, as $\ln\frac{q}{p} = \ln\left(1+\frac{q-p}{p}\right) \leqq \frac{q-p}{p}$, we also have (using (39)) $2p\log_2\frac{q}{p} \leqq 2\log_2 e\cdot(q-p) = \log_2 e\cdot\sum_{x\in X}|p(x)-q(x)|$.

In view of these estimates, (42) and (37) imply (38).

1. Introduction

Intuitively, a channel is a device capable of transmitting information. We restrict our attention to discrete channels, where information transmission is achieved by sending message symbols chosen from a finite or countably infinite set Y at consecutive time points. As a rule, the message symbols are subject to random distortions - generally referred to as noise - thus the set $\tilde{Y}$ of possible received symbols may be different from Y The sets Y and $\tilde{Y}$ are referred to as input and output alphabets of the channel; the elements of Y and $\tilde{Y}$ are often called input and output letters, respectively. In a discrete channel also the output alphabet $\tilde{Y}$ is assumed to be at most countable.

Mathematically, the channel is characterized by its input and output alphabets, as well as the probabilistic laws describing the noise, i. e. the distortion of the message symbols in course of the transmission.

The simplest discrete channel is the binary symmetric channel, abbreviated as BSC. Its input and output alphabets consist of the binary digits 0 and 1 , i. e. $Y = \tilde{Y} = \{0,1\}$ and the noise is characterized by the probability p that a symbol 0 is incorrectly received as 1 or that a symbol 1 is incorrectly received as 0 ; these probabilities are supposed to be equal. Moreover, the BSC is defined to be a memoryless channel, i. e. if any sequence of binary digits is transmitted (at time instants $t = 1, 2, \ldots$, say), the noise is supposed to affect these message symbols independently. This model can be used for many practical problems, at least as a first approximation. In practice, however, the errors often have a tendency to occur in groups, i. e., a received digit is more likely to be in error after an incorrectly received digit than after a correctly received one. If this tendency to "error bursts" is not negligible, a more complex model is needed and one will be facing a channel with memory.

Another simple channel model is the binary erasure channel, abbreviated as BEC. Here again $Y = \{0,1\}$ but it is assumed that neither digit can be incorrectly received as another digit, while it may happen that the received digit cannot be identified; the probability p of the latter event is the same for both possible input symbols. Also the BEC is, by definition, a memoryless channel.

Channel coding theory deals with the possibilities of reliable transmission of messages through noisy channels. Intituively, these messages may be thought of as resulting from a previous source coding, e. g. they may be binary sequences reliably representing blocks of k consecutive outputs of an information source which is required to be transmitted over the channel. Then, by varying the source block length k we have a certain liberty in choosing the number N of possible messages. If $N = 2^{nR}$, the possible messages may be interpreted as binary sequences of length nR. This means that, on the average, R binary digits are to be transmitted per one channel use; for this reason, $R = \frac{1}{n} \log_2 N$ is called the <u>rate</u> of the transmission.

An essential point is that in channel coding theory no probability distribution governing the selection of the message to be transmitted is assumed to be given. This means that channel coding should be independent of the statistical properties of the source from which the messages arise; this is important not only because in this way both mathematical theory and implementation become simpler but also because in practice the statistical properties of the source are rarely known sufficiently well. The starting point of our investigations will be the following model.

Messages from a finite set Z of size N are to be transmitted by n channel uses. This means that <u>code words</u> of length n i.e. sequences of n symbols of the input alphabet Y of the channel $v = v(z)$ are to be associated to each possible message $z \in Z$; the message $z \in Z$ is transmitted by sending the code-word $v = v(z)$.

The code words should be selected so that on the basis of the received sequence $\tilde{v}$ the code word sent (hence also the message sent) can be reconstructed with a small probability of error.

If there were no noise, all n - sequences (sequences of length n) of input symbols could be used as code words. In the presence of noise, however, the possibility of reliable decoding can be expected only if the code words are "far from each other", thus, N must be considerable less than s^n , where s stands for the size of the input alphabet of the channel.

There are two - related - theoretical problems in this respect :

(i) if we fix a bound λ of the probability of erroneous decoding we are ready to tolerate, determine the maximum $N(n,\lambda)$ of the size N of the set of messages transmissible by n channel uses, or, at least, determine the asymptotic behaviour of $N(n,\lambda)$ as $n \to \infty$;

(ii) if messages from a set of size $N = 2^{nR}$ are to be transmitted by n channel uses, determine the asymptotic behaviour of the least upper bound of the probability of error (if $n \to \infty$) if optimal encoding and decoding rules are used.

The typical solution of these problems is

that $N(n, \lambda)$ approximately equals 2^{nC} where C is a constant called the capacity of the channel and the bound of error probability converges exponentially to 0 provided that the rate R is less than the capacity C.

In the sequel we shall be dealing with problems (i) and (ii), with the aim of giving rigorous formulations to the loose statements of the last paragraph and proving them for at least certain classes of noisy channels.

We do not enter, however, into the problem of how to construct and how to implement optimal or nearly optimal encoding and decoding schemes; in this respect let us refer to the lectures of Professor Berlekamp. In spite of the fact that the theory to be developed does not give rise to concrete code constructions, its results will be not only of theoretical but also of considerable practical interest. In fact, the knowledge of the optimal performance a given channel is theoretically capable of, enables one to draw important practical conclusions; in particular, to evaluate the efficiency of concrete encoding and decoding schemes.

2. Mathematical Definition of Discrete Communication Channels with Noise.

Before giving a mathematical definition of a noisy communication channel, let us consider first a simpler object, to be called observation channel.

Suppose that V and $\tilde{V}$ are two finite or countably infinite sets such that if an element $v \in V$ is selected, there corresponds a $\tilde{v} \in \tilde{V}$ to it; the connection of v and $\tilde{v}$, however, need not be deterministic, rather, $\tilde{v}$ is selected according to a PD $\mathcal{P}_v = \{p_v(\tilde{v}), \tilde{v} \in \tilde{V}\}$. Suppose further that $v \in V$ is the object of our interest but only $\tilde{v} \in \tilde{V}$ is accessible to observation; in this case one may say that information on $v \in V$ is available only indirectly, through the observation channel $(V, \tilde{V}, p)$.

<u>Definition 2.1.</u> An <u>observation channel</u> $(V, \tilde{V}, p)$ is defined by the set of PD's $\mathcal{P}_v = \{p_v(\tilde{v}), \tilde{v} \in \tilde{V}\}$ on $\tilde{V}$ where v runs over V

<u>Definition 2.2.</u> If $v_1, \ldots, v_N$ are certain elements of V to which there exist pairwise disjoint subsets $B_1, \ldots, B_N$

of $\tilde{V}$ such that

$$p_{v_i}(B_i) \geq 1 - \lambda \qquad i = 1, \ldots, N \tag{2.1}$$

where $0 < \lambda < 1$ and

$$p_v(B) \overset{\text{def}}{=} \sum_{\tilde{v} \in B} p_v(\tilde{v}) \qquad (v \in V, B \subset \tilde{V}), \tag{2.2}$$

the elements $v_1, \ldots, v_N$ are called λ - <u>distinguishable</u> by the observation channel $(V, \tilde{V}, p)$.

The condition (2.1) means that if v is known to have been selected from the v_i's and if, in case $\tilde{v} \in B_i$, the decision $v = v_i$ is taken, this decision will be correct with a probability of error at most λ, no matter which of the v_i's has been selected.

A noisy communication channel with input alphabet Y and output alphabet $\tilde{Y}$ can be characterized by giving for all possible "word lengths" $n = 1, 2, \ldots$ and for all possible n - sequences $v = y_1 \ldots y_n$ of letters sent at consecutive time instants $t = 1, \ldots, n$ the probabilities $p_v(\tilde{v})$ of receiving different n - sequences $\tilde{v} = \tilde{y}_1 \ldots \tilde{y}_n$ $(y_i \in Y, \tilde{y}_i \in \tilde{Y}, i = 1, \ldots, n)$.

<u>Definition 2.3.</u> A <u>simple communication channel</u> with input alphabet Y and output alphabet $\tilde{Y}$ is defined as a sequence of observation channels $(Y^n, \tilde{Y}^n, p)$, $n = 1, 2, \ldots$,

where Y^n ($\tilde{Y}^n$) denotes the set of all possible n -sequences of letters of Y ($\tilde{Y}$). The probabilities $p_v(\tilde{v})$ ($v \in Y^n$, $\tilde{v} \in \tilde{Y}^n$) are called the n -<u>dimensional transition probabilities</u> of the channel, and the observation channel $(Y^n, \tilde{Y}^n, p)$ will be referred to as the n -dimensional observation channel.

Of course, in order to get proper mathematical models of physical channels, some consistency conditions should be imposed on the n - dimensional transition probabilities.

In accordance with the intuitive concept given in sect. 1, a channel is called memoryless if the n - dimensional transition probabilities are just products of the one-dimensional ones, i. e. if

$$p_v(\tilde{v}) = \prod_{i=1}^{n} p_{y_i}(\tilde{y}_i) \quad \text{if} \quad v = y_1 \ldots y_n,\ \tilde{v} = \tilde{y}_1 \ldots \tilde{y}_n . \tag{2.3}$$

To be more exact, equation (2.3) defines the <u>stationary</u> memoryless channels; since non-stationary memoryless channels are of little practical interest, they will not be considered here.

Memoryless channels provide a reasonable first approximation for many physical channels. Often, however, there are non-negligible after-effects in the channel and a second approximation taking into account the existence of memory is needed.

A model of channels with memory sufficient for almost all purposes is the finite-state channel, where a parameter called the state of the channel capable of a finite number of different values comprises the memory on the previously sent and received symbols.

A finite-state channel with input alphabet Y, output alphabet $\tilde{Y}$ and set of states A (where A is a finite set) is defined by the set of "transition probabilities"

$$p_{y,a}(\tilde{y},b) \; ; \; (y \in Y, \tilde{y} \in \tilde{Y}, a, b \in A)$$

where $p_{y,a}(\tilde{y},b)$ is the probability that if a letter $y \in Y$ is sent when the state is $a \in A$ the letter $\tilde{y} \in \tilde{Y}$ will be received and the state changes to $b \in A$. The probability that when sending an n - sequence $v = y_1 \ldots y_n$ with initial state $a_0 \in A$, the received n - sequence will be $\tilde{v} = \tilde{y}_1 \ldots \tilde{y}_n$ and the sequence of new states at the instants $t = 1, \ldots, n$ will be $c = a_1 \ldots a_n$ is defined as

$$p_{v,a_0}(\tilde{v},c) = \prod_{i=1}^{n} p_{y_i,a_{i-1}}(\tilde{y}_i,a_i). \tag{2.4}$$

The formal correspondence with definition 2.3 is established by defining the n - dimensional transition probabilities as

$$p_{v}(\tilde{v}) = \sum_{c \in A^n} p_{v,a_0}(\tilde{v}, c) \tag{2.5}$$

where A^n means the set of all possible n-sequences of elements of A.

Observe that (2.5) involves the assumption that the initial state (the state at $t = 0$) is known. If this is not the case, one actually has a finite number of different simple channels in the sense of definition 2.3, one for each possible initial state.

In practice the situation often arises that the performance of the channel depends on some unknown parameters and the method of coding and decoding should be designed so as to ensure reliable transmission regardless of the actual value of the unknown parameters. In such cases one speaks of a compound channel.

Definition 2.4. A compound channel is a set of simple communication channels. These simple channels will be called the components of the compound channel.

Remark. The parameter that specifies the component simple channels is often called the state of the compound channel. This term should not be confused with the state of a finite-state channel. The "state" of a compound channel plays the role of the initial state of a finite-state channel, and remains constant during the transmission.

The model of channel coding introduced in sect. 1 suggests to define the maximum size $N(n, \lambda)$ of the set of messages transmissible by n channel uses over a simple communication channel (for a fixed error probability bound λ $(0 < \lambda < 1)$) as the maximum number of λ - distinguishable elements for the n - dimensional observation channel $(Y^n, \tilde{Y}^n, p)$.

Definition 2.5. For a simple channel, $N(n, \lambda)$ is defined as the largest N for which there exists N code - words $v_1, \ldots, v_N$ of length n (i. e. $v_i \in Y^n, i = 1, \ldots, N$) and corresponding disjoint decoding sets $B_i \subset \tilde{Y}_n$ $(i = 1, \ldots, N)$ satisfying the inequalities (2.1.). For a compound channel, $N(n, \lambda)$ is defined similarly, being the conditions $p_{v_i}(B_i) \geq 1 - \lambda$ $(i = 1, \ldots, N)$ interpreted so that the inequalities hold for the n - dimensional transition probabilities of each component simple channel.

Then $\log_2 N(n, \lambda)$ is a reasonable measure of the amount of information (in bits) reliably transmissible by n channel uses, where the value of λ specifies the reliability requirement.

For this reason

$$(2.6) \qquad C(\lambda) = \overline{\lim_{n\to\infty}} \frac{1}{n} \log_2 N(n,\lambda) \qquad (0 < \lambda < 1)$$

may be called the λ - capacity of the channel.

Of course, the limit

$$(2.7) \qquad C = \lim_{\lambda\to 0} C(\lambda)$$

is of major importance. This is the least upper bound of the rates at which it is possible to transmit with arbitrarily small error probability bound $\lambda > 0$. In fact, if $R < C$, then, $R < C(\lambda)$ for any λ $(0 < \lambda < 1)$, hence there exist arbitrarily large numbers n such that $2^{nR} < N(n,\lambda)$, thus, 2^{nR} messages are transmissible within the error probability bound λ ; if, however, $R > C$, then $R > C(\lambda)$ for some $\lambda > 0$,hence $2^{nR} > N(n,\lambda)$ for sufficiently large n , thus 2^{nR} messages are not transmissible within this bound λ .

It is by no means obvious, rather, an important result of information theory that $C > 0$ for all "reasonable" channels. This means, that one can overcome the noise by coding and transmit as reliably as required without paying for increasing reliability by decreasing rate (provided that $R < C$). The only price for this is that proper codes of sufficiently long word length should be used (*)

It is an interesting feature that for wide classes of communication channels the "λ-capacity" $C(\lambda)$ does not depend on λ $(0<\lambda<1)$. In this case one may assert that C is the least upper bound of the rates at which it is possible to reliably transmit, without specifying the reliability requirement, i.e. the value of λ.

Definition 2.6. The number C defined for an arbitrary communication channel by (2.6.) and (2.7.) is called the channel capacity. If $C(\lambda)$ on (2.6.) does not depend λ $(0<\lambda<1)$, one says that capacity exists in the strong sense and C is called strong capacity.

There is another approach to the concept of channel capacity.

For an observation channel $(V, \tilde{V}, p)$ suppose that η is a RV taking values in V with distribution $\mathcal{P} = \{p(v),\ v \in V\}$. Then, there is an associated RV $\tilde{\eta}$ with values in $\tilde{V}$, such that for every $v \in V$

$$P\{\tilde{\eta} = \tilde{v} \mid \eta = v\} = p_v(\tilde{v}) \qquad (\text{if } p(v) > 0). \tag{2.8}$$

(*) This may be an expensive price, as regards implementation, since "good" codes are usually very complex ones.

The mutual information

$$(2.9) \qquad I(\eta \wedge \tilde{\eta}) = \sum_{v \in V, \tilde{v} \in \tilde{V}} p(v)\, p_v(\tilde{v}) \log_2 \frac{p_v(\tilde{v})}{q_{\mathcal{P}}(\tilde{v})} \overset{def}{=} f(\mathcal{P})$$

where

$$(2.10) \qquad q_{\mathcal{P}}(\tilde{v}) \overset{def}{=} \sum_{v \in V} p(v)\, p_v(\tilde{v})$$

is a measure of the(average) amount of information on η obtained from observing $\tilde{\eta}$.

<u>Definition 2.7.</u> The least upper bound of $I(\eta \wedge \tilde{\eta}) = f(\mathcal{P})$ with respect to the PD $\mathcal{P}$, i.e.

$$(2.11) \qquad C^I \overset{def}{=} \sup_{\mathcal{P}} \sum_{v \in V, \tilde{v} \in \tilde{V}} p(v)\, p_v(\tilde{v}) \log_2 \frac{p_v(\tilde{v})}{q_{\mathcal{P}}(\tilde{v})}$$

is called the <u>information capacity</u> of the given observation channel.

Intuitively, information capacity is the maximum amount of information obtainable over the observation channel $(V, \tilde{V}, p)$.

If either of the sets V and $\tilde{V}$ is finite, C^I is certainly finite, namely (*)

(*) The estimates follow from $I(\eta \wedge \hat{\eta}) \leq H(\eta)$ and $I(\eta \wedge \hat{\eta}) \leq H(\tilde{\eta})$.

$$C^I \leq \log_2 |V|, \quad C^I \leq \log_2 |\tilde{V}|, \tag{2.12}$$

respectively, where $| \ |$ denotes the number of elements of the set considered. If both V and $\tilde{V}$ are infinite, C^I may or may not be finite.

For a simple communication channel let $C^I(n)$ denote the information capacity of the n - dimensional observation channel, i. e.

$$C^I(n) = \sup_{\mathcal{P}} \sum_{\substack{v \in Y^n \\ \tilde{v} \in \tilde{Y}^n}} p(v)\, p_v(\tilde{v}) \log_2 \frac{p_v(\tilde{v})}{q_{\mathcal{P}}(\tilde{v})} \tag{2.13}$$

where $\mathcal{P} = \{p(v),\ v \in Y^n\}$ runs over all PD's on Y^n and $q_{\mathcal{P}}(\tilde{v})$ is defined by (2. 10), with V replaced by Y^n . In case of compound channels the mutual information on the right hand side of (2. 13) depends (for a fixed PD $\mathcal{P}$) on the state of the channel, i. e. on the particular component simple channel that has been selected. Since this is unknown, it seems reasonable to take the infimum of this mutual information and to define

$$C^I(n) = \sup_{\mathcal{P}} \left\{ \inf \sum_{v \in Y^n} \sum_{\tilde{v} \in \tilde{Y}^n} p(v)\, p(\tilde{v}) \log_2 \frac{p_v(\tilde{v})}{q_{\mathcal{P}}(\tilde{v})} \right\} \tag{2.14}$$

where the infimum is taken with respect to the components

of the compound channel.

Definition 2.8. The information capacity of a communication channel is defined as

(2.15) $$C^I = \lim_{n\to\infty} \frac{1}{n} C^I(n)$$

(provided that the limit exists), where $C^I(n)$ is defined by (2.13) or (2.14), respectively.

Observe that if either of the alphabets Y and $\tilde{Y}$ is finite then so is C^I (if it exists). In fact, (2.12) and (2.15) imply

(2.16) $$C^I \leq \log_2 |Y| , \quad C^I \leq \log_2 |\tilde{Y}| .$$

One of the main results of information theory - called the coding theorem or Shannon's theorem - is that for sufficiently wide classes of channels capacity (as defined by (2.7.)) and information capacity (as defined by (2.15)) are equal. The statement that for any ε and λ ($\varepsilon > 0$, $0 < \lambda < 1$) the inequality $N(n,\lambda) \geq 2^{n(C^I-\varepsilon)}$ holds if n is large enough (implying $C \geq C^I$) is referred to as the direct part of the coding theorem, while the statement that for any $\varepsilon > 0$ there exists $\lambda > 0$ such that $N(n,\lambda) < 2^n(C^I+\varepsilon)$ for sufficiently large n (implying $C \leq C^I$)

is the converse part(*)of the coding theorem.

It often happens that the last inequality holds for arbitrary ε and λ $(\varepsilon > 0,\ 0 < \lambda < 1)$, if n is sufficiently large. In that case one says that the strong converse holds, since this - together with the direct part of the coding theorem - implies the existence of channel capacity in the strong sense.

Remark. The terms coding theorem and weak (strong) converse are sometimes used in a broader sense which, however, will not be needed in the sequel.

3. Coding Theorems for Noisy Channels.

Let us show first that the (weak) converse of the coding theorem holds for arbitrary (simple or compound) communication channels.

Lemma 3.1. (Fano inequality). Let $(V, \tilde{V}, p)$ be an observation channel and let η and $\tilde{\eta}$ be RV's taking values in V and $\tilde{V}$, respectively, such that (2.8) is valid

(*) Or the weak converse, when compared with the strong converse described in the next paragraph.

and

$$(3.1) \qquad P\{\eta = v_i\} = \frac{1}{N} \qquad (i = 1, \ldots, N)$$

where $v_1, \ldots, v_N$ are some fixed elements of V. Suppose that there exists a function f with a domain $\tilde{V}$ such that

$$(3.2) \qquad P\{f(\tilde{\eta}) = \eta\} \geq 1 - \lambda \qquad (0 < \lambda < 1).$$

Then,

$$(3.3) \qquad \log_2 N \leq \frac{I(\eta \wedge \hat{\eta}) + \bar{h}(\lambda)}{1 - \lambda}$$

where

$$(3.4) \qquad \bar{h}(\lambda) = \begin{cases} -\lambda \log_2 \lambda - (1-\lambda) \log_2 (1-\lambda) & \text{if } \lambda \leq \frac{1}{2} \\ 1 & \text{if } \lambda > \frac{1}{2}. \end{cases}$$

<u>Proof.</u> Let χ denote the RV taking the value 1 if $f(\tilde{\eta}) = \eta$ and 0 if $f(\tilde{\eta}) \neq \eta$ Then,

$$(3.5) \qquad H(\eta|\tilde{\eta}) \leq H(\eta, \chi|\tilde{\eta}) = H(\eta|\chi, \tilde{\eta}) + H(\chi|\tilde{\eta}).$$

Since for $\chi = 1$ and $\tilde{\eta}$ fixed, η can take on only one value (namely $f(\tilde{\eta})$) while for $\chi = 0$ and $\tilde{\eta}$ fixed η can take on $N-1$ different values, we have (by inequality (30) of the Preliminaries)

$$H(\eta|\tilde{\eta},\chi) \leq P\{\chi=0\}\log_2(N-1) \leq \lambda \log_2(N-1). \qquad (3.6)$$

Further, on account of (3.2) and (3.4)

$$H(\chi|\tilde{\eta}) \leq H(\chi) = -P\{\chi=0\}\log_2 P\{\chi=0\} - P\{\chi=1\}\log_2 P\{\chi=1\} \leq \bar{h}(\lambda). \qquad (3.7)$$

Since $I(\eta\wedge\tilde{\eta}) = H(\eta) - H(\eta|\tilde{\eta})$, (3.1) and the estimates (3.5), (3.6), (3.7) give rise to

$$I(\eta\wedge\tilde{\eta}) \geq \log_2 N - \lambda \log_2(N-1) - \bar{h}(\lambda) \qquad (3.8)$$

whence (3.3) immediately follows.

Theorem 3.1. For arbitrary (simple or compound) communication channels

$$\log_2 N(n,\lambda) \leq \frac{C^I(n)+\bar{h}(\lambda)}{1-\lambda} \qquad (n=1,2,\ldots,0<\lambda<1); \qquad (3.9)$$

this implies

$$C(\lambda) \leq \frac{C^I}{1-\lambda} \qquad \text{and} \qquad C \leq C^I \qquad (3.10)$$

provided that C^I exists.

Proof. Let $v_1, \ldots, v_n$ be code words of length n and let $B_1, \ldots, B_N$ be decoding sets

$$(v_i \in Y^n,\ B_i \subset \tilde{Y}^n,\ i=1,\ldots,N,\ B_i \cap B_j = \emptyset \text{ if } i \neq j)$$

such that $p_{v_i}(B_i) \geq 1-\lambda \qquad i = 1, \ldots, N$; if a compound channel is considered, the latter inequalities should hold for each component simple channel, cf. definition 2.4.

Thus, lemma 3.1 can be applied for the n-dimensional observation channel $(Y^n, \tilde{Y}^n, p)$ with f being defined so that $f(\tilde{v}) = v_i$ if $\tilde{v} \in B_i$. Using the expression (2.9) for the mutual information $I(\eta \wedge \tilde{\eta})$, we obtain

$$\text{(3.11)} \qquad \log_2 N \leq \frac{1}{1-\lambda}\left(\sum_{v \in V^n}\sum_{\tilde{v} \in \tilde{V}^n} p(v)\, p_v(\tilde{v}) \log_2 \frac{p_v(\tilde{v})}{q_p(\tilde{v})} + \bar{h}(\lambda)\right)$$

where $p(v) = \frac{1}{N}$ if $v = v_i \quad (i = 1, \ldots, N)$ and $p(v) = 0$ else. In case of a compound channel, inequality (3.11) holds for each component simple channel, thus, the double sum in (3.11) may be replaced by its infimum with respect to the components.

On account of (2.13) and (2.14), hence immediately follows (3.9), since, according to definition 2.5, one may assume that $N = N(n, \lambda)$.

The remaining statements of theorem 3.1 are obvious corollaries of this result, according to (2.6),

(2.15) and (2.7).

Our next aim is to prove the direct part of the coding theorem for the class of memoryless channels.

Definition 3.1. A simple channel is memoryless if its n-dimensional transition probabilities are products of the one-dimensional ones as in (2.3). A compound channel is memoryless if its components are memoryless simple channels.

The idea of proving $N(n, \lambda) > 2^{n(C^I - \varepsilon)}$, i.e. the existence of $N > 2^{n(C^I - \varepsilon)}$ code-words $v_i \in Y^n$ and corresponding decoding sets $B_i \in \tilde{Y}^n$ satisfying the inequalities (2.1) will be to pick N code-words at random (according to a properly chosen PD on Y^n) and then check that with the B_i 's chosen by the maximum likelihood principle, the resulting code is a good one, with a large probability (or, rather, it can be easily modified to yield a good code). This method, referred to as the random coding technique, is widely used in information theory and deserves particular attention.

The following lemma 3.2 comprises the main idea of the proof. Let $(V, \tilde{V}, p)$ be an observation channel, $v_1, \ldots, v_N$ be certain elements of V, and $B_i = B_i(v_1, \ldots, v_N) \subset \tilde{V}$ be the set of those $\tilde{v} \in \tilde{V}$ for which

$$p_{v_i}(\tilde{v}) > p_{v_j}(\tilde{v}) \quad \text{for all } j \ (1 \leq j \leq N,\ j \neq i). \qquad (3.12)$$

Obviously, the sets B_i are disjoint.

Now, let the elements $v_1, \ldots, v_N$ be picked at random; then, also the sets B_i as well as the probabilities $p_{v_i}(B_i)$ will depend on chance.

Lemma 3.2. Let $\eta_1, \ldots, \eta_N$ be pairwise independent RV's taking values in V with the same PD $\mathcal{P} = \{p(v), \; v \in V\}$. Write

$$\lambda_i = p_{\eta_i}\left(B_i^c(\eta_1, \ldots, \eta_N)\right); \quad \bar{\lambda} = \frac{1}{N} \sum_{i=1}^{N} \lambda_i \tag{3.13}$$

Then, setting

$$W = \left\{ (v, \tilde{v}) : \frac{p_v(\tilde{v})}{q_p(\tilde{v})} \leq \alpha N \right\} \quad (\alpha > 1) \tag{3.14}$$

and

$$P(W) = \sum_{(v,\tilde{v}) \in W} p(v)\, p_v(\tilde{v}) \tag{3.15}$$

we have (*)

$$E\bar{\lambda} < \frac{1}{\alpha} + P(W). \tag{3.16}$$

(*) $q_p(\tilde{v})$ is defined as in (2.10), i.e. $q_p(\tilde{v}) = \sum_{v \in V} p(v)\, p_v(\tilde{v})$.

Corollary. If $\frac{1}{\alpha} + P(W) \leq \frac{1}{2}$, there exist N different elements $v_i \in V$ and N disjoint sets $B_i \subset \tilde{V}$ such that

$$p_{v_i}(B_i^c) < 2\left(\frac{2}{\alpha} + P(W)\right) \qquad i = 1, \ldots, N. \tag{3.17}$$

Proof. (3.13) may be written as

$$\lambda_i = \sum_{\tilde{v} \in \tilde{V}} p_{\eta_i}(\tilde{v})\, \chi_i(\tilde{v}) \tag{3.18}$$

where $\chi_i(\tilde{v}) = 0$ if $\tilde{v} \in B_i(\eta_1, \ldots, \eta_N)$ and $\chi_i(\tilde{v}) = 1$ if $\tilde{v} \notin B_i(\eta_1, \ldots, \eta_N)$.

The conditional expectation of λ_i under the condition $\eta_i = v$ equals - using the fact that $\tilde{v} \notin B_i(\eta_1, \ldots, \eta_N)$ iff $p_{\eta_i}(\tilde{v}) \leq p_{\eta_j}(\tilde{v})$ for at least one $j \neq i$, cf. (3.12) -

$$E(\lambda_i \mid \eta_i = v) = \sum_{v \in \tilde{V}} p_v(\tilde{v})\, E(\chi_i(\tilde{v}) \mid \eta_i = v) =$$

$$= \sum_{\tilde{v} \in \tilde{V}} p_v(\tilde{v})\, P\{p_v(\tilde{v}) \leq p_{\eta_j}(\tilde{v}) \text{ for at least one } j \neq i \mid \eta_i = v\}. \tag{3.19}$$

Since the η_j's are pairwise independent and

identically distributed, we have

$$(3.20)\quad P\{p_v(\tilde{v}) \leq p_{\eta_j}(\tilde{v}) \text{ for at least one } j \neq i \mid \eta_i = v\} \leq$$

$$\leq \sum_{\substack{j=1\\ j\neq i}}^{N} P\{p_v(\tilde{v}) \leq p_{\eta_j}(\tilde{v}) \mid \eta_i = v\} =$$

$$= \sum_{\substack{j=1\\ j\neq i}}^{N} P\{p_v(\tilde{v}) \leq p_{\eta_j}(\tilde{v})\} =$$

$$= (N-1)\, P\{p_v(\tilde{v}) \leq p_{\eta_1}(\tilde{v})\}.$$

By using this estimate for $(v, \tilde{v}) \notin W$ and the trivial one that a probability is ≤ 1 for $(v, \tilde{v}) \in W$, we obtain from (3.19)

$$E\lambda_i = \sum_{v\in V} p(v) E(\lambda_i \mid \eta_i = v) \leq (N-1) \sum_{(v,\tilde{v})\notin W} p(v) p_v(\tilde{v})\, P\{p_v(\tilde{v}) \leq$$

$$(3.21)\qquad \leq p_{\eta_1}(\tilde{v})\} + \sum_{(v,\tilde{v})\in W} p(v)\, p_v(\tilde{v}).$$

Here in the first sum

$$P\{p_v(\tilde{v}) \leq p_{\eta_1}(\tilde{v})\} = \sum_{v' : p_v(\tilde{v}) \leq p_{v'}(\tilde{v})} p(v') \leq$$

$$(3.22)\qquad \leq \sum_{v'\in V} \frac{p(v')\, p_{v'}(\tilde{v})}{\alpha N q_p(\tilde{v})} = \frac{1}{\alpha N}$$

using that in force of (3.14) $\dfrac{p_v(\tilde{v})}{\alpha N q_p(\tilde{v})} > 1$ if $(v, \tilde{v}) \notin W$, whence even more $\dfrac{p_{v'}(\tilde{v})}{\alpha N q_p(\tilde{v})} > 1$

if $p_{v'}(\tilde{v}) \leqq p_v(\tilde{v})$.

From (3. 21), (3. 22) and (3. 15) follows

$$E\lambda_i \leq \frac{N-1}{\alpha N} \sum_{(v,\tilde{v}) \notin W} p(v)\, p_v(\tilde{v}) + P(W) < \frac{1}{\alpha} + P(W). \quad (3.23)$$

This inequality holds for $i = 1, \ldots, N$, thus, (3. 16) is proved.

To prove the corollary, substitute $\frac{\alpha}{2}$ instead of α and $2N$ instead of N (this does not affect (3. 14)), yielding

$$E\left\{\frac{1}{2N} \sum_{i=1}^{2N} p_{\eta_i}\left(B_i^c(\eta_1, \ldots, \eta_{2N})\right)\right\} < \frac{2}{\alpha} + P(W).$$

Hence obviously follows that for some realization

$$(v_1, \ldots, v_{2N}) \text{ of } (\eta_1, \ldots, \eta_{2N})$$

$$\frac{1}{2N} \sum_{i=1}^{2N} p_{v_i}\left(B_i^c(v_1, \ldots, v_{2N})\right) < \frac{2}{\alpha} + P(W), \quad (3.24)$$

and then, for at least N indices i, the inequality (3. 17) must hold. In force of the assumption $\frac{2}{\alpha} + P(W) \leqq \frac{1}{2}$, these v_i 's must be all different, since $v_i = v_j$ for some $j \neq i$ would imply $B_i = \emptyset$, cf. (3. 12).

<u>Lemma 3.3.</u> In case of (simple or compound) memoryless channels

$$(3.25) \qquad C^I(n) = n\, C^I(1) \,, \qquad C^I = C^I(1).$$

<u>Proof</u>. If $\mathcal{P} = \left\{ p(v) \,,\ v \in Y^n \right\}$ is an arbitrary PD on Y^n consider the PD $\mathcal{Q}_\mathcal{P} = \left\{ q_\mathcal{P}(\tilde{v}),\ \tilde{v} \in \tilde{Y}^n \right\}$ where

$$(3.26) \qquad q_\mathcal{P}(\tilde{v}) = \sum_{v \in Y^n} p(v)\, p_v(\tilde{v}),$$

as well as the marginal distributions $\mathcal{P}^i$ and $\mathcal{Q}^i_\mathcal{P}$ on Y and $\tilde{Y}$ respectively, defined by

$$(3.27) \qquad p^i(y) = \sum_{\substack{v = y_1 \ldots y_n \in Y^n \\ y_i = y}} p(v) \,; \quad q^i_\mathcal{P}(\tilde{y}) = \sum_{\substack{\tilde{v} = \tilde{y}_1 \ldots \tilde{y}_n \in \tilde{Y}^n \\ \tilde{y}_i = \tilde{y}}} q_\mathcal{P}(\tilde{v})$$
$$(i = 1, \ldots, n) \,.$$

Of course, for compound channels the PD $\mathcal{Q}_\mathcal{P}$ and its marginals depend on which component of the compound channel has been selected.

Observe that the assumption (2.3) implies

$$(3.28) \qquad q^i_\mathcal{P}(\tilde{y}) = \sum_{y \in Y} p^i(y)\, p_y(\tilde{y})$$

and also that in case of $p(v) = \prod_{i=1} p(y_i) \quad (v = y_1 \ldots y_n)$

there holds

$$q_{\mathcal{P}}(\tilde{v}) = q^*_{\mathcal{P}}(\tilde{v}) \overset{def}{=} \prod_{i=1}^{n} q^i_{\mathcal{P}}(\tilde{y}_i) \qquad (\tilde{v} = \tilde{y}_1 \dots \tilde{y}_n) \tag{3.29}$$

as well.

In force of (2. 3), we have for arbitrary $\mathcal{P} = \{ p(v),\ v \in Y^n \}$

$$\log_2 \frac{p_v(\tilde{v})}{q_{\mathcal{P}}(\tilde{v})} = \sum_{i=1}^{n} \log_2 \frac{p_{y_i}(\tilde{y}_i)}{q^i_{\mathcal{P}}(\tilde{y}_i)} + \log_2 \frac{q^*_{\mathcal{P}}(\tilde{v})}{q_{\mathcal{P}}(\tilde{v})} \ . \tag{3.30}$$

Thus, (2. 13) reduces - by using (2. 3) and (3. 26) - to

$$C^I(n) = \sup_{\mathcal{P}} \left\{ \sum_{i=1}^{n} \sum_{y_i \in Y} \sum_{\tilde{y}_i \in \tilde{Y}} p^i(y_i)\, p_{y_i}(\tilde{y}_i) \log_2 \frac{p_{y_i}(\tilde{y}_i)}{q^i_{\mathcal{P}}(\tilde{y}_i)} + \right.$$

$$\left. + \sum_{\tilde{v} \in \tilde{V}} q_{\mathcal{P}}(\tilde{v}) \log_2 \frac{q^*_{\mathcal{P}}(\tilde{v})}{q_{\mathcal{P}}(\tilde{v})} \right\} . \tag{3.31a}$$

In case of a compound channel the only modification to be made is - cf. (2. 14) - to take infimum with in the braces with respect to the component simple channels of the compound channel.

On the right hand side of (3. 31a) the first term depends on $\mathcal{P}$ through the marginal distributions $\mathcal{P}^i$

only (see(3.28)) while the second term equals

$$- I\left(\mathfrak{Q}_{\mathcal{P}} \,\|\, \mathfrak{Q}^*_{\mathcal{P}}\right),$$

which is ≤ 0 with equality only if $\mathfrak{Q}_{\mathcal{P}} = \mathfrak{Q}^*_{\mathcal{P}}$; for the latter $p(v) = \prod_{i=1}^{n} p^i(y_i) \quad (v = y_1 \ldots y_n)$ is a sufficient condition, cf. (3.29). Thus, (3.31a) implies (*)

$$(3.31b) \quad C^I(n) = n \sup_{\mathcal{P}} \left\{ \sum_{y \in Y} \sum_{\tilde{y} \in \tilde{Y}} p(y) p_y(\tilde{y}) \log_2 \frac{p_y(\tilde{y})}{q_{\mathcal{P}}(\tilde{y})} \right\} = n\, C^I(1),$$

completing the proof.

<u>Theorem 3.2.</u> Let us be given a memoryless simple channel with finite information capacity. Then, for any λ and ε $(0 < \lambda < 1,\ \varepsilon > 0)$ there holds

$$(3.32) \qquad N(n, \lambda) > 2^{n(C^I - \varepsilon)}$$

if n is sufficiently large.

If the input and output alphabets Y and $\tilde{Y}$ are finite, (3.32) may be strengthened to

$$(3.33) \qquad N(n, \lambda) > 2^{n(C^I - K n^{-1/2})} \qquad (n = 1, 2, \ldots)$$

(*) In (3.31b), $\mathcal{P}$ runs over the PD's on the set Y ; in case of a compound channel, an "inf" should be inserted within the braces.

where K depends on λ but not on n

Proof. Let $\eta_1, \eta_2, \ldots$ be a sequence of independent RV's taking values in Y with the same PD

$$\mathcal{P} = \{p(y), y \in Y\}.$$

Consider the associated RV's $\tilde{\eta}_1, \tilde{\eta}_2, \ldots$ taking values in $\tilde{Y}$, where

$$P\{\tilde{\eta}_1 = \tilde{y}_1, \ldots, \tilde{\eta}_n = \tilde{y}_n \mid \eta_1 = y_1, \ldots, \eta_n = y_n\} = \prod_{i=1}^{n} p_{y_i}(\tilde{y}_i).$$

Then, the pairs $(\eta_1, \tilde{\eta}_1), (\eta_2, \tilde{\eta}_2), \ldots$ are mutually independent and we have

$$\iota_{\eta_1, \ldots \eta_n \wedge \tilde{\eta}_1, \ldots, \tilde{\eta}_n} = \sum_{i=1}^{n} \iota_{\eta_i \wedge \tilde{\eta}_i}, \tag{3. 34}$$

where the information densities $\iota_{\eta_i \wedge \tilde{\eta}_i}$ are mutually independent RV's. The weak law of large numbers (cf. e. g. Feller [3] , p. 228) implies that $\frac{1}{n} \sum_{i=1}^{n} \iota_{\eta_i \wedge \tilde{\eta}_i}$ converges in probability to

$$E\, \iota_{\eta_1 \wedge \tilde{\eta}_1} = I(\eta_1 \wedge \tilde{\eta}_1),$$

i. e.

$$P\left\{\left|\frac{1}{n} \iota_{\eta_1, \ldots, \eta_n \wedge \tilde{\eta}_1, \ldots, \tilde{\eta}_n} - I(\eta_1 \wedge \tilde{\eta}_1)\right| > \varepsilon'\right\} \to 0 \tag{3. 35}$$

as $n \to \infty$ for any $\varepsilon' > 0$.

Consider the n - dimensional observation channel $(Y^n, \tilde{Y}^n, p)$ with the joint distributions of $\eta_1, \dots, \eta_n$ and of $\tilde{\eta}_1, \dots, \tilde{\eta}_n$ playing the roles of $\mathcal{P}$ and $\mathcal{Q}_{\mathcal{P}}$, respectively. In this case

(3.36) $$\iota_{\eta_1,\dots,\eta_n \wedge \tilde{\eta}_1,\dots,\tilde{\eta}_n} = \log_2 \frac{p_v(\tilde{v})}{q_{\mathcal{P}}(\tilde{v})} \text{ if } \eta_1 \dots \eta_n = v,\ \tilde{\eta}_1 \dots \tilde{\eta} = \tilde{v},$$

implying

(3.37) $$P(W) = P\left\{ \iota_{\eta_1,\dots,\eta_n \wedge \tilde{\eta}_1,\dots,\tilde{\eta}_n} \le \log_2 \alpha N \right\}$$

in lemma 3.2.

Suppose that the PD $\mathcal{P}$ has been chosen so that $I(\eta_1 \wedge \tilde{\eta}_1) > C^I - \frac{\varepsilon}{2}$, say, and set $N = 2^{n(C^I - \varepsilon)}$; then, from (3.35) and (3.37), follows $P(W) < \frac{\lambda}{4}$ if n is large enough, and the corollary of lemma 3.2 implies, taking $\alpha = \frac{8}{\lambda}$, that for n large enough there exist $N = 2^{n(C^I - \varepsilon)}$ different sequences $v_i \in Y^n$ to which disjoint sets $B_i \subset \tilde{Y}^n$ with $p_{v_i}(B_i^c) < \lambda$ may be chosen.

If the alphabets Y and $\tilde{Y}$ are finite sets, the sum

$$f(\mathcal{P}) = \sum_{y \in Y} \sum_{\tilde{y} \in \tilde{Y}} p(y)\, p_v(y) \log_2 \frac{p_v(y)}{q_{\mathcal{P}}(y)}$$

is a continuous function of $\mathcal{P} = \{p(v), v \in Y\}$, hence its-maximum is attained. Let the maximizing $\mathcal{P}$ be the common distribution of the RV's η_i; then, $I(\eta_1 \wedge \tilde{\eta}_1) = C^I$. The finiteness of Y and $\tilde{Y}$ implies that the information densities $\iota_{\eta_i \wedge \tilde{\eta}_i}$ have finite variance σ^2, say. Then, the expectation and variance of $\iota_{\eta_1,\dots,\eta_n \wedge \tilde{\eta}_1,\dots,\tilde{\eta}_n}$ equals $n C^I$ and $n\sigma^2$, respectively, cf. (3.34), thus, setting $N = 2^{n(C^I - Kn^{-1/2})}$, from (3.37) follows by Chebyshev's inequality

$$P(W) \leq P\left\{\left|\iota_{\eta_1,\dots,\eta_n \wedge \tilde{\eta}_1,\dots,\tilde{\eta}_n} - nC^I\right| \geq K\sqrt{n} - \log_2 \alpha\right\} \leq$$

$$\leq \frac{n\sigma^2}{(K\sqrt{n} - \log_2\alpha)^2}. \tag{3.38}$$

Taking again $\alpha = \frac{8}{\lambda}$ and choosing K so large that the bound in (3.38) be $\leq \frac{\lambda}{4}$, the bound of (3.17) becomes $\leq \lambda$, completing the proof.

<u>Example.</u> Memoryless simple channels with finite alphabets are most conveniently specified by their one-dimensional transition probabilities arranged into an s by $\tilde{s}$ matrix π where s and $\tilde{s}$ denote the number of elements of Y and $\tilde{Y}$, respectively. The channel is called <u>symmetric,</u> if $\tilde{Y}$ can be partitioned into disjoint subsets $\tilde{Y}_1, \dots, \tilde{Y}_k$ (of $\tilde{s}_1, \dots, \tilde{s}_k$ elements, say; $k = 1$ is permitted) so that each of the corresponding $s \times \tilde{s}_i$ submatrices of the transition probability matrix π has the property

that its rows as well as its columns are permutations of each other. E.g., the BSC and BEC described in the introduction, characterized by the matrices $\pi = \begin{pmatrix} 1-p & p \\ p & 1-p \end{pmatrix}$ and $\pi = \begin{pmatrix} 1-p & 0 & p \\ 0 & 1-p & p \end{pmatrix}$, respectively, are symmetric channels. We show that the capacity of symmetric channels is attained for uniform "input distribution"; i.e. for η and $\tilde{\eta}$ taking values in Y and $\tilde{Y}$, respectively, and satisfying $P\{\tilde{\eta} = \tilde{y} \mid \eta = y\} = p_y(\tilde{y})$, $I(\eta \wedge \hat{\eta})$ is maximized if the distribution of η is $\mathcal{P}_0 = \left(\frac{1}{s}, \ldots, \frac{1}{s}\right)$. Since for a symmetric channel the conditional entropy $H(\tilde{\eta} \mid \eta)$ does not depend on $\mathcal{P}$ (the distribution of η),

$I(\eta \wedge \hat{\eta}) = H(\tilde{\eta}) - H(\tilde{\eta} \mid \eta)$ is maximized by maximizing $H(\tilde{\eta}) = H(\mathcal{Q}_{\mathcal{P}})$. From the symmetry assumption follows that the sums $\sum_{\tilde{y} \in \tilde{Y}_i} q_{\mathcal{P}}(\tilde{y})$ do not depend on $\mathcal{P}$, and also that $q_{\mathcal{P}_0}(\tilde{y})$ is constant for $\tilde{y} \in \tilde{Y}_i$ $(i = 1, \ldots, k)$.

Hence,

$$H(\mathcal{Q}_{\mathcal{P}_0}) - H(\mathcal{Q}_{\mathcal{P}}) \overset{\text{def}}{=} -\sum_{\tilde{y} \in \tilde{Y}} q_{\mathcal{P}_0}(\tilde{y}) \log_2 q_{\mathcal{P}_0}(\tilde{y}) + \sum_{\tilde{y} \in \tilde{Y}} q_{\mathcal{P}}(\tilde{y}) \log_2 q_{\mathcal{P}}(\tilde{y}) =$$

$$= \sum_{\tilde{y} \in \tilde{Y}} q_{\mathcal{P}}(\tilde{y}) \log_2 \frac{q_{\mathcal{P}}(\tilde{y})}{q_{\mathcal{P}_0}(\tilde{y})} = I(\mathcal{Q}_{\mathcal{P}} \| \mathcal{Q}_{\mathcal{P}_0}) \geq 0,$$

proving our assertion. In particular, the capacity of a BSC is $C = C^I = 1 + p \log_2 p + (1-p) \log_2 (1-p)$ and that of a BEC turns out to be $C = C^I = p$.

Theorem 3.3. The result (3.33) is valid for compound memoryless channels with finite input and output alphabets, as well.

Proof. Suppose first that the compound channel has only a finite number of components, say t . Let the probabilities referring to different component simple channels be specified by upper indices, and consider the auxiliary observation channel $(Y^n, \tilde{Y}^n, \bar{p})$ where

$$\bar{p}_v(\tilde{v}) = \frac{1}{t} \sum_{k=1}^{t} p_v^k(\tilde{v}) \qquad (v \in Y^n,\ \tilde{v} \in \tilde{Y}^n). \tag{3.39}$$

Observe that, if $v_1, \ldots, v_N$ are $\frac{\lambda}{t}$-distinguishable by $(Y^n, \tilde{Y}^n, \bar{p})$, i.e. if

$$\bar{p}_{v_i}(B_i) \stackrel{def}{=} \frac{1}{t} \sum_{k=1}^{t} p_{v_i}^k(B_i) \geq 1 - \frac{\lambda}{t}, \quad i = 1, \ldots, N \tag{3.40}$$

for certain disjoint sets $B_i \subset \tilde{Y}^n$, then $p_{v_i}^k(B_i) \geq 1 - \lambda$ holds for all i and k implying $N(n, \lambda) \geqq N$, cf. definition 2.5.

We wish to apply the corollary of lemma 3.2 to the auxiliary observation channel $(Y^n, \tilde{Y}^n, \bar{p})$. Then, the role of W is played by

$$W = \left\{(v, \tilde{v}) : \frac{\bar{p}_v(\tilde{v})}{\bar{q}_p(\tilde{v})} \leq \alpha N \right\} \subset Y^n \times \tilde{Y}^n \tag{3.41}$$

where

$$\bar{q}_p(\tilde{v}) \overset{\text{def}}{=} \sum_{v \in Y^n} p(v)\bar{p}_v(\tilde{v}) = \frac{1}{t}\sum_{k=1}^{t} q_p^k(\tilde{v}); \text{ we set}$$

$$(3.42) \qquad W_k = \left\{(v,\tilde{v}) : \frac{p_v^k(\tilde{v})}{q_p^k(\tilde{v})} \le \beta N\right\} \subset Y^n \times \tilde{Y}^n .$$

For $(v,\tilde{v}) \in W \setminus W_k$ we have

$$q_p^k(\tilde{v}) < \frac{p_v^k(\tilde{v})}{\beta N} \le \frac{t}{\beta N}\,\bar{p}_v(\tilde{v}) \le$$

$$\le \frac{t}{\beta N}\cdot \alpha N \bar{q}_p(\tilde{v}) = \frac{\alpha}{\beta}\sum_{j=1}^{t} q_p^j(\tilde{v}).$$

From the previous results, the following inequalities can therefore be obtained

$$p^k(W \setminus W_k) \overset{\text{def}}{=} \sum_{(v,\tilde{v}) \in W \setminus W_k} p(v)\,p_v^k(\tilde{v}) \le \sum_{\substack{\tilde{v}\,:\,(v,\tilde{v}) \in W \setminus W_k \\ \text{for some } v \in Y^n}} q_p^k(\tilde{v}) \le$$

$$(3.43) \qquad \le \frac{\alpha}{\beta}\sum_{j=1}^{t}\sum_{\tilde{v} \in \tilde{Y}^n} q_p^j(\tilde{v}) = \frac{\alpha}{\beta}\, t \,.$$

Thus, $\bar{P}(W) \overset{\text{def}}{=} \sum_{(v,\tilde{v}) \in W} p(v)\,\bar{p}_v(\tilde{v})$ may be bounded as

$$\bar{P}(W) = \frac{1}{t}\sum_{k=1}^{t} P^k(W) \leq \frac{1}{t}\sum_{k=1}^{t}\left(\frac{\alpha}{\beta}t + P^k(W_k)\right) =$$

$$= \frac{\alpha}{\beta}t + \frac{1}{t}\sum_{k=1}^{t} P^k(W^k). \tag{3.44}$$

Now, (3. 33) follows similarly as in the proof of theorem 3. 2 :

Let $\eta_1, \eta_2, \ldots$ be independent RV's with values in Y, with the same distribution $\mathcal{P} = \{p(y), y \in Y\}$ chosen such as to maximize

$$\inf_{1\leq k\leq t} \sum_{y\in Y}\sum_{\tilde{y}\in\tilde{Y}} p(y)\, p_y^k(\tilde{y}) \log_2 \frac{p_y^k(\tilde{y})}{q_{\mathcal{P}}^k(\tilde{y})} ;$$

then, the latter quantity equals $C^I(1) = C^I$. Let $\tilde{\eta}_1^k, \tilde{\eta}_2^k, \ldots$ be RV's with values in $\tilde{Y}$ satisfying

$$P\left\{\tilde{\eta}_1^k = y_1, \ldots, \tilde{\eta}_n^k = y_n \mid \eta_1 = y_1, \ldots, \eta_n = y_n\right\} = \prod_{i=1}^{n} p_{y_i}^k(\tilde{y}_i).$$

We apply the corollary of lemma 3. 2 for the observation channel $(Y^n, \tilde{Y}^n, \bar{p})$ with the joint distribution of $\eta_1, \ldots, \eta_n$ playing the role of $\mathcal{P}$ and with $N = 2^{n(C^I - k n^{-1/2})}$. Taking into account (3. 44), where - to the analogy of (3. 38) -

$$P^k(W_k) = P\left\{ \iota_{\eta_1,\dots,\eta_n \wedge \tilde{\eta}_1^k,\dots,\tilde{\eta}_n^k} \leq \log_2 \beta + nC^I - K\sqrt{n} \right\} \leq$$

$$\leq P\left\{ \left| \iota_{\eta_1,\dots,\eta_n \wedge \tilde{\eta}_1^k,\dots,\tilde{\eta}_n^k} - nI(\eta_1 \wedge \tilde{\eta}_1^k) \right| \geq K\sqrt{n} - \log_2 \beta \right\} \leq$$

$$\leq \frac{\sigma_k^2}{(K\sqrt{n} - \log_2 \beta)^2} \leq \frac{\lambda}{8t} \tag{3.45}$$

(if K has been chosen sufficiently large) we obtain, by setting $\alpha = \frac{8t}{\lambda}$, $\beta = \frac{t^3}{\lambda^2}$, the bound $\frac{\lambda}{t}$ on the right hand side of (3.17). Thus, there exist $N = 2^{n(C^I - Kn^{-1/2})}$ sequences $v_1, \dots, v_N$ in Y^n which are $\frac{\lambda}{t}$-distinguishable by the auxiliary observation channel $(Y^n, \tilde{Y}^n, \bar{p})$ completing the proof of (3.33) (for the case of a finite number of component simple channels).

The general case can be settled by an approximation argument. For given n, let us approximate the transition probability matrices of the component channels by auxiliary ones, with entries of form $k \cdot 2^{-[\sqrt{n}]}$ (k integer), such that the approximation be elementwise, within accuracy $2^{-[\sqrt{n}]}$ The approximating matrices define an auxiliary compound channel (depending on n) with a finite number t_n of component simple channels, where $t_n < (2^{\sqrt{n}})^{s\tilde{s}}$ (s and $\tilde{s}$ denote the size of Y and $\tilde{Y}$, respectively). For this auxiliary channel the above argument applies, since the dependence of t on n - as long as $\log_2 \beta$ (with $\beta = \frac{64t^3}{\lambda^2}$ as above) remains of order at most $\sqrt{n}$ does not affect its validity, cf. (3. 45). It is easy to check that the differences between probabilities of form $p_v(B)$ $(v \in Y^n,\ B \subset \tilde{Y}^n)$ for the components of the original channel on one hand and for the corresponding components of the auxiliary channel on the other hand, are negligible and so is the difference in C^I between the two channels (the details are left to the reader). Thus, the validity of (3. 33) for the auxiliary channel implies the same for the original compound channel, as well.

The results proved so far imply that for (simple or compound)(*)memoryless channels capacity equals information capacity, and it can be obtained by maximizing a function of reasonable complexity of $P = \{p(y),\ y \in Y\}$. Also more than this is true; memoryless channels (at least those with finite alphabets) have capacity in the strong sense, i. e.

$$\lim_{n \to \infty} \frac{1}{n} \log_2 N(n, \lambda) = C, \tag{3.46}$$

for arbitrary λ $(0 < \lambda < 1)$. This means that transmission at a rate above capacity cannot be reliable, no matter how mild reliability requirement is used (if it is formulated in terms of an error probability bound).

(*) For compound channels, this has been proved only in the case of finite alphabets.

Let us send forward a lemma.

Lemma 3.4. For an observation channel $(V, \tilde{V}, p)$ where V and $\tilde{V}$ are finite sets, the function (2.9) of $\mathcal{P} = \{p(v), v \in V\}$ is maximized iff for some constant C

$$(3.47) \qquad I(\mathcal{P}_v \| \mathcal{Q}_{\mathcal{P}}) \overset{\text{def}}{=} \sum_{\tilde{v} \in \tilde{V}} p_v(\tilde{v}) \log_2 \frac{p_v(\tilde{v})}{q_{\mathcal{P}}(\tilde{v})} \leq C$$

for all $v \in V$, with equality if $p(v) > 0$. Then, the constant C equals the information capacity C^I and for any PD $\mathcal{Q} = \{q(\tilde{v}), \tilde{v} \in \tilde{V}\} \neq \mathcal{Q}_{\mathcal{P}}$ we have

$$\max_{v \in V} I(\mathcal{P}_v \| \mathcal{Q}) \overset{\text{def}}{=}$$

$$(3.48) \qquad \overset{\text{def}}{=} \max_{v \in V} \sum_{\tilde{v} \in V} p_v(\tilde{v}) \log_2 \frac{p_v(\tilde{v})}{q(\tilde{v})} > C^I.$$

Proof. Since V and $\tilde{V}$ are finite sets,

$$f(\mathcal{P}) = \sum_{v \in V} \sum_{\tilde{v} \in \tilde{V}} p(v) p_v(\tilde{v}) \log_2 \frac{p_v(\tilde{v})}{q_{\mathcal{P}}(\tilde{v})} \quad \text{with} \quad q_{\mathcal{P}}(\tilde{v}) = \sum_{v \in V} p(v) p_v(\tilde{v})$$

is a continuous function of $\mathcal{P}$, hence its maximum is attained. Let $\mathcal{P} = \{p(v), v \in V\}$ be such a PD for which the maximum C^I is attained. For fixed $v_1 \in V$ with $p(v_1) > 0$ and $v_2 \in V$, consider the PD's $\mathcal{P}_t$ obtained from $\mathcal{P}$ by subtracting t from $p(v_1)$ and adding t to $p(v_2)$; then, the function $f(\mathcal{P}_t)$ of t $(0 \leq t \leq p(v_1))$ is maximized for $t = 0$, i.e. its (right) derivative must be ≤ 0 . Carrying out differentiation :

$$0 \geq \frac{df(\mathcal{P}_t)}{dt}\bigg|_{t=0} = \sum_{\tilde{v}\in\tilde{V}} \Big(- p_{v_1}(\tilde{v}) \log_2 \frac{p_{v_1}(\tilde{v})}{q_{\mathcal{P}}(\tilde{v})} +$$

$$+ p_{v_2}(\tilde{v}) \log \frac{p_{v_2}(\tilde{v})}{q_{\mathcal{P}}(\tilde{v})} - \sum_{v\in V} p(v)\, p_v(\tilde{v}) \frac{-p_{v_1}(\tilde{v}) + p_{v_2}(\tilde{v})}{q_{\mathcal{P}}(\tilde{v})} \log_2 e\Big) =$$

$$= -I(\mathcal{P}_{v_1} \| \mathcal{Q}_{\mathcal{P}}) + I(\mathcal{P}_{v_2} \| \mathcal{Q}_{\mathcal{P}}). \qquad (3.49)$$

Thus, we have $I(\mathcal{P}_{v_1} \| \mathcal{Q}_{\mathcal{P}}) \geq I(\mathcal{P}_{v_2} \| \mathcal{Q}_{\mathcal{P}})$ whenever $p(v_1) > 0$, proving the necessity of (3.47). Suppose now that $\mathcal{Q} = \{ q(\tilde{v}), \tilde{v} \in \tilde{V} \}$ is a PD on $\tilde{V}$ satisfying

$$I(\mathcal{P}_v \| \mathcal{Q}) \overset{\text{def}}{=} \sum_{\tilde{v}\in\tilde{V}} p_v(\tilde{v}) \log_2 \frac{p_v(\tilde{v})}{q(\tilde{v})} \leq C^I \qquad (3.50)$$

for all $v \in V$. Multiplying by $p(v)$ and summing for $v \in V$, in view of the identity $\log_2 \frac{p_v(\tilde{v})}{q(\tilde{v})} = \log_2 \frac{p_v(\tilde{v})}{q_{\mathcal{P}}(\tilde{v})} + \log_2 \frac{q_{\mathcal{P}}(\tilde{v})}{q(\tilde{v})}$ we obtain

$$\sum_{v\in V} \sum_{\tilde{v}\in\tilde{V}} p(v)\, p_v(\tilde{v}) \log_2 \frac{p_v(\tilde{v})}{q_{\mathcal{P}}(\tilde{v})} + \sum_{v\in V} \sum_{\tilde{v}\in\tilde{V}} p(v)\, p_v(\tilde{v}) \log_2 \frac{q_{\mathcal{P}}(\tilde{v})}{q(\tilde{v})} \overset{\text{def}}{=}$$

$$\overset{\text{def}}{=} f(\mathcal{P}) + I(\mathcal{Q}_{\mathcal{P}} \| \mathcal{Q}) \leq C^I . \qquad (3.51)$$

Since $f(\mathcal{P}) = C^I$, this is a contradiction unless $\mathfrak{Q} = \mathfrak{Q}_\mathcal{P}$, proving (3.48).

It follows, in particular, that if (3.47) is fulfilled for some $\mathcal{P} = \{p(v), v \in V\}$ and C, then, necessarily, $C \geq C^I$. On the other hand, multiplying (3.47) by $p(v)$ and summing for all $v \in V$ we obtain $f(\mathcal{P}) = C$, proving both $C = C^I$ and the sufficiency of the conditions (3.47).

<u>Remark.</u> Lemma 3.4. has an interesting "geometric" interpretation: looking at the I - divergence as an information-theoretic measure of "distance" of PD's, (3.47) and (3.48) mean that the information capacity C^I is the "radius" of the set of PD's $\mathcal{P}_v$ on $\tilde{V}$ and $\mathfrak{Q} = \mathfrak{Q}_\mathcal{P}$ is the "centre" of this set (though the maximizing $\mathcal{P}$ need not be unique, $\mathfrak{Q}_\mathcal{P}$ is uniquely determined by lemma 3.4).

Sometimes lemma 3.4 may be used to calculate information capacity. If V and $\tilde{V}$ are of the same size, (3.37) - supposing that the equalities hold - may be solved for $q_\mathcal{P}(\tilde{v})$ and C (unless the matrix $(p_v(\tilde{v}))_{\substack{v \in V \\ \tilde{v} \in \tilde{V}}}$ is singular); then, if the equations $\sum_{v \in V} p(v)\, p_v(\tilde{v}) = q_\mathcal{P}(\tilde{v})$ have a solution $\mathcal{P} = \{p(v), v \in V\}$ which is a PD, the obtained C is the information capacity. The practical value of this method should not to be overestimated; to calculate the numerical value of information capacity, convex programming methods are often preferable (the function to be maximized is concave, by theorem 4 of the preliminaries).

Theorem 3.4. For (simple) memoryless channels with finite alphabets we have for arbitrary λ $(0 < \lambda < 1)$

$$N(n,\lambda) < 2^{n(C' + Kn^{-1/2})} \qquad (n = 1,2,\ldots) \qquad (3.52)$$

where K depends on λ but not on n .

Proof. First we show for an arbitrary observation channel $(V, \tilde{V}, p)$ that if there exists a PD $\mathcal{Q} = \{q(\tilde{v}), \tilde{v} \in \tilde{V}\}$ on $\tilde{V}$ and an $\alpha > 0$ such that for every $v \in V$

$$p_v(A_v) > \alpha + \lambda\,, \quad A_v \overset{\text{def}}{=} \left\{ \tilde{v} : \frac{p_v(\tilde{v})}{q(\tilde{v})} \le \alpha N \right\} \qquad (3.53)$$

then, the maximal number of λ - distinguishable elements of V is less than N

In fact, suppose that there exist elements $v_i \in V$ and disjoint sets $B_i \subset \tilde{V}$ $(i = 1, \ldots, N)$ such that $p_{v_i}(B_i) \ge 1 - \lambda$.

Then,

$$\alpha N \sum_{\tilde{v} \in B_i \cap A_{v_i}} q(\tilde{v}) \geq \sum_{\tilde{v} \in B_i \cap A_{v_i}} p_{v_i}(\tilde{v}) \geq$$

(3.54) $$\geq p_{v_i}(A_{v_i}) - p_{v_i}(B_i^c) > \alpha$$

whence, summing for $i = 1, \ldots, N$, we get the contradiction $\alpha N > \alpha N$.

Consider now the n - dimensional observation channel $(Y^n, \tilde{Y}^n, p)$ of the given memoryless channel. Let $\{q_0(\tilde{y}), \ \tilde{y} \in \tilde{Y}\}$ be the PD $\mathcal{Q}_{\mathcal{P}}$ corresponding to a PD $\mathcal{P} = \{p(y), \ y \in Y\}$ for which the maximum C^I of $\sum_{y \in Y} \sum_{\tilde{y} \in \tilde{Y}} p(y)\, p_y(\tilde{y}) \log_2 \frac{p_y(\tilde{y})}{q_{\mathcal{P}}(\tilde{y})}$ is attained, and set

(3.55) $$q(\tilde{v}) = \prod_{i=1}^{n} q_0(\tilde{y}_i) \quad \text{if} \quad \tilde{v} = \tilde{y}_1 \ldots \tilde{y}_n .$$

Let $0 < \lambda < 1$, $\alpha > 0$ be fixed so that $\alpha + \lambda < 1$, and take $N = 2^{n(C^I + kn^{-1/2})}$. Then, A_v of (3.53) may be written for $v = y_1 \ldots y_n$ as

(3.56) $$A_v = \left\{ \tilde{y}_1 \ldots \tilde{y}_n : \sum_{i=1}^{n} \log_2 \frac{p_{y_i}(\tilde{y}_i)}{q_0(\tilde{y}_i)} \leq \log_2 \alpha + nC^I + k\sqrt{n} \right\}.$$

For $v = y_1 \ldots y_n$ fixed, the terms $\log_2 \frac{p_{y_i}(\tilde{y}_i)}{q_0(\tilde{y}_i)}$ may be regarded as mutually independent RV's with respect to (*) the PD $\mathcal{P}_v$ on $\tilde{Y}^n$. Their expectation is, in

---

(*) More formally: the probability space $(\Omega, \mathcal{F}, P)$ is defined by $\Omega = \tilde{Y}_n$, $\mathcal{F}$ = the σ - algebra of all subsets of $\tilde{Y}^n$, $P(B) = p_v(B)$ for $B \subset \tilde{Y}^n$.

force of lemma 3.4

$$E \log_2 \frac{p_{y_i}(\tilde{y}_i)}{q_0(\tilde{y}_i)} \overset{\text{def}}{=} \sum_{\tilde{y}_i \in \tilde{Y}} p_{y_i}(\tilde{y}_i) \log_2 \frac{p_{y_i}(\tilde{y}_i)}{q_0(\tilde{y}_i)} \leq C^I , \tag{3.57}$$

while their variance is bounded by some constant M depending on the channel only (and not on v).

Thus, from Chebyshev's inequality follows that

$$p_v(A_v^c) = P\left\{ \sum_{i=1}^{n} \log_2 \frac{p_{y_i}(\tilde{y}_i)}{q_0(\tilde{y}_i)} > \log_2 \alpha + nC^I + K\sqrt{n} \right\} \leq$$

$$\leq P\left\{ \left| \sum_{i=1}^{n} \left(\log_2 \frac{p_{y_i}(\tilde{y}_i)}{q_0(\tilde{y}_i)} - E \log_2 \frac{p_{y_i}(\tilde{y}_i)}{q_0(\tilde{y}_i)} \right) \right| > \log_2 \alpha + K\sqrt{n} \right\} \leq$$

$$\leq \frac{nM}{(\log_2 \alpha + K\sqrt{n})^2} < 1 - (\alpha + \lambda) \tag{3.58}$$

for all $v \in Y^n$ provided that K has been chosen sufficiently large, i.e.; (3.53) is satisfied and the maximal number $N(n, \lambda)$ of λ - distinguishable elements of Y^n is less than $2^{n(C^I + Kn^{-\frac{1}{2}})}$.

The essential point of theorem 3.4 is that for memoryless channels with finite alphabets the strong converse of the coding theorem is valid. As a matter of mathematical interest let us mention that for memoryless channels with infinite alphabets the strong converse may not hold. For compound memoryless channels with finite alphabets the statement of theorem 3.4 remains valid, but the proof is more complex, thus, we omit it.

On inspecting the proof of theorem 3.2 it turns out that the only role of the assumption of having a channel without memory was to ensure that the right hand side of (3.37) approaches 0 if $N = 2^{n(C^I - \varepsilon)}$ and $n \to \infty$.

Thus, the direct part of the coding theorem has been proved, eventually, for a much broader class of channels.

Definition 3.2. A simple channel with finite information capacity $C^I > 0$ is said to be information stable if for any $\varepsilon > 0$ and $\delta > 0$ there exist RV's $\eta = (\eta_1, \dots, \eta_n)$ and $\tilde{\eta} = (\tilde{\eta}_1, \dots, \tilde{\eta}_n)$ connected by (2.8) and satisfying

$$(3.59) \qquad P\left\{\left|\frac{1}{n}\, \iota_{\eta_1,\dots,\eta_n \wedge \tilde{\eta}_1,\dots,\tilde{\eta}_n} - C^I\right| > \varepsilon\right\} < \delta ,$$

if n is sufficiently large.

By the above proof, the information stability of a simple channel is a sufficient condition for the validity of the direct part of the coding theorem; it is not difficult to show that this condition is necessary, as well.

Unfortunately, the usefulness of this necessary and sufficient condition is limited, since for channels with memory information stability is often difficult to check.

We mention, without proof, an important class of information stable channels, the so-called indecomposable finite state channels.

For a finite state channel with initial state $a_0 \in A$ consider the probability $p_{v,a_0}(a_n)$ that after transmitting a sequence $v = y_1 \dots y_n \in Y^n$ the state will be $a_n \in A$. According to (2.4), $p_{v,a_0}(a_n)$ may be defined as the sum of the probabilities $p_{v,a_0}(\tilde{v}, c) = \prod_{i=1}^{n} p_{y_i, a_{i-1}}(\tilde{y}_i, a_i)$: summing for all $\tilde{v} = \tilde{y}_1 \dots \tilde{y}_n \in \tilde{Y}^n$ and for all $c = a_1 \dots a_n \in A^n$ where the final state a_n is the given one.

<u>Definition 3.3.</u> A finite state channel is <u>indecomposable</u> if for any $\varepsilon > 0$ there exists $n_0 = n_0(\varepsilon)$ such that for any choice of the states a_0, a_0' and a_n and of the sequence $v \in Y^n$,

$$\left| p_{v,a_0}(a_n) - p_{v,a_0'}(a_n) \right| < \varepsilon \tag{3.60}$$

provided that $n \geqq n_0$.

Intuitively, a finite state channel is indecomposable if the effect of the initial state dies away with time.

A necessary and sufficient condition for a finite state channel to be indecomposable is that for some fixed n and some $a_n \in A$ the probabilities $p_{v,a_0}(a_n)$ be positive for all possible choices of the initial state a_0

and the transmitted sequence $v \in Y^n$, where the state a_n may depend on v but not on a_0.

An indecomposable channel is always information stable, hence its capacity equals the information capacity; the latter may be shown to exist and to be independent of the initial state a_0

The proof of these statements will be omitted. We remark that, for such channels, no satisfactory method is known to calculate information capacity; in fact, its numerical value is unknown already for very simple indecomposable finite state channels.

4. The Reliability Function of a Communication Channel.

In Section 3 we have considered the problem (i) of channel coding, posed in Section 1; now we turn to problem (ii).

<u>Definition 4.1.</u> For a given communication channel and $R>0$ let $\lambda(n, R)$ denote the infimum of the numbers $\lambda>0$ for which there exist $N = 2^{n.R}$ code words of length n $v_i \in Y^n$ and (disjoint) decoding sets $B_i \subset \tilde{Y}^n$ with $p_{v_i}(B_i) \geqq 1-\lambda$, $i = 1, \dots, N$; in case of a compound channel the latter inequalities

should hold for each component.

From definitions 2.5 and 4.1 follows that $\lambda(n, R) \to 0$ as $n \to \infty$ for any fixed $R < C$. The typical situation is that $\lambda(n, R)$ decreases exponentially; this motivates

Definition 4.2. The function

$$E(R) = \overline{\lim_{n \to \infty}} \left(-\frac{1}{n} \log_2 \lambda(n, R) \right) \tag{4.1}$$

is called the reliability function(*) of the channel.

The significance of this function is obvious. When we want to transmit at a fixed rate R, the knowledge of the value of $E(R)$ is even more important than that of the channel capacity.

Of course, for practical purposes estimates of $\lambda(n, R)$ valid for all n (rather than asymptotic ones) are preferable, particularly if one wants to use codes of moderate word length.

We shall need the following modification of lemma 3.2 and its corollary.

Lemma 4.1. With the notations of lemma 3.2, we have

(*) Also the term "error exponent" is used.

for $\frac{1}{2} \leq \alpha \leq 1$

$$(4.2) \qquad E\,\bar{\lambda}\,(\eta_1,\ldots,\eta_N) \leq (N-1)^{\frac{1-\alpha}{\alpha}} \sum_{\tilde{v}\in\tilde{V}} \Big(\sum_{v\in V} p(v)\, p_v^{\alpha}(\tilde{v})\Big)^{\frac{1}{\alpha}}.$$

Corollary. There exist N different elements $v_1, \ldots, v_N$ of V and disjoint subsets $B_1, \ldots, B_N$ of $\tilde{V}$ such that

$$(4.3) \qquad p_{v_i}(B_i^c) \leq 2\,(2N-1)^{\frac{1-\alpha}{\alpha}} \sum_{\tilde{v}\in\tilde{V}} \Big(\sum_{v\in V} p(v)\, p_v^{\alpha}(\tilde{v})\Big)^{\frac{1}{\alpha}}, \quad i=1,\ldots,N$$

provided that the right hand side is less than 1 .

Proof. Observe that from (3.20) follows

$$P\left\{ p_v(\tilde{v}) \leq p_{\eta_j}(\tilde{v}) \text{ for at least one } j \neq i \,\middle|\, \eta_i = v \right\} \leq$$

$$(4.4) \qquad \leq \Big((N-1)\, P\left\{ p_v(\tilde{v}) \leq p_{\eta_1}(\tilde{v}) \right\} \Big)^{\rho}$$

as well, if $0 \leq \rho \leq 1$. In fact, if the right hand side of (3.20) is less than 1 , it will be increased by raising to the power ρ, while else (4.4) is trivial. We also have

$$P\left\{ p_v(\tilde{v}) \leq p_{\eta_1}(\tilde{v}) \right\} =$$

$$(4.5) \qquad = \sum_{v':\, p_v(\tilde{v}) \leq p_{v'}(\tilde{v})} p(v') \leq \sum_{v'\in V} p(v) \left(\frac{p_{v'}(\tilde{v})}{p_v(\tilde{v})} \right)^{\alpha}$$

for arbitrary $\alpha > 0$.

From (3.19), (4.4) and (4.5) follows

$$E\lambda_i = \sum_{v \in V} p(v)\, E(\lambda_i \mid \eta_i = v) \leqq$$

$$\leqq \sum_{v \in V} \sum_{\tilde{v} \in \tilde{V}} p(v)\, p_v(\tilde{v}) \left[(N-1) \sum_{v' \in V} p(v') \left(\frac{p_{v'}(\tilde{v})}{p_v(\tilde{v})} \right)^{\alpha} \right]^{\rho} =$$

$$= (N-1)^{\rho} \sum_{\tilde{v} \in \tilde{V}} \left(\sum_{v \in V} p(v)\, p_v^{1-\rho\alpha}(\tilde{v}) \right) \left(\sum_{v \in V} p(v)\, p_v^{\alpha}(\tilde{v}) \right)^{\rho}. \tag{4.6}$$

Taking $\rho = \frac{1}{\alpha} - 1$ (this is where the assumption $\frac{1}{2} \leqq \alpha \leqq 1$ is used), the right hand side of (4.6) reduces to that of (4.2). Thus, being $\bar{\lambda}(\eta_1, \ldots, \eta_N) = \frac{1}{N} \sum_{i=1}^{N} \lambda_i$ (4.2) is proved.

To prove the corollary, substitute $2N$ for N and observe that (4.2) implies that for some realization $(v_1, \ldots, v_{2N})$ of $(\eta_1, \ldots, \eta_{2N})$

$$\frac{1}{N} \sum_{i=1}^{2N} p_{v_i}\left(B_i^c(v_1, \ldots, v_{2N})\right) \leqq$$
$$\leqq (2N-1)^{\frac{1-\alpha}{\alpha}} \sum_{\tilde{v} \in \tilde{V}} \left(\sum_{v \in V} p(v)\, p_v^{\alpha}(\tilde{v}) \right)^{\frac{1}{\alpha}}. \tag{4.7}$$

Hence, for at least N indices i, the inequality (4.3) must hold. That these v_i 's are all different, follows in the same way as in the corollary of lemma 3.2.

To obtain the best bound,

$$g_\alpha(\mathcal{P}) \stackrel{\text{def}}{=} \sum_{\tilde{v} \in \tilde{V}} \left(\sum_{v \in V} p(v)\, p_v^{\alpha}(\tilde{v}) \right)^{\frac{1}{\alpha}} \tag{4.8}$$

should be minimized as $\mathcal{P}$ runs over the PD's $\mathcal{P} = \{p(v),\ v \in V\}$.

It will be convenient to consider also

$$f_\alpha(\mathcal{P}) \overset{\text{def}}{=} \frac{\alpha}{\alpha-1} \log_2 g_\alpha(\mathcal{P}) \tag{4.9}$$

Define, for $0 < \alpha < 1$

$$\begin{aligned} G(\alpha) &= \inf_{\mathcal{P}} g_\alpha(\mathcal{P}); \\ F(\alpha) &= \sup_{\mathcal{P}} f_\alpha(\mathcal{P}) = \frac{\alpha}{\alpha-1} \log_2 G(\alpha). \end{aligned} \tag{4.10}$$

<u>Lemma 4.2.</u> For $0 < \alpha < 1$ we have $0 < g_\alpha(\mathcal{P}) \leq 1$, $0 \leq f_\alpha(\mathcal{P}) < \infty$; $g_\alpha(\mathcal{P})$ and $f_\alpha(\mathcal{P})$ are increasing functions of α (for fixed $\mathcal{P}$) and

$$\lim_{\alpha \to 1-0} f_\alpha(\mathcal{P}) = \log_2 e \cdot \left.\frac{\partial g_\alpha(\mathcal{P})}{\partial \alpha}\right|_{\alpha=1} =$$

$$= f(\mathcal{P}) \overset{\text{def}}{=} \sum_{v \in V} \sum_{\tilde{v} \in \tilde{V}} p(v)\, p_v(\tilde{v}) \log_2 \frac{p_v(\tilde{v})}{q_{\mathcal{P}}(\tilde{v})}, \tag{4.11}$$

where $\left.\frac{\partial g_\alpha(\mathcal{P})}{\partial \alpha}\right|_{\alpha=1}$ means a left derivative and $q_{\mathcal{P}}(\tilde{v}) = \sum_{v \in V} p(v)\, p_v(\tilde{v})$. If $f(\mathcal{P}) > 0$ then $g_\alpha(\mathcal{P})$ and $f_\alpha(\mathcal{P})$ are strictly increasing functions of α.

In particular, $G(\alpha)$ and $F(\alpha)$ are increasing functions of α and

$$\lim_{\alpha \to 1-0} F(\alpha) = C^{I} \tag{4.12}$$

where $C^{I} = \sup_{\mathcal{P}} f(\mathcal{P})$ is the information capacity.

If V and $\tilde{V}$ are finite sets and $0 < \alpha < 1$ is fixed, $g_\alpha(\mathcal{P})$ is minimized ($f_\alpha(\mathcal{P})$ maximized) if for some constant K

$$\sum_{\tilde{v} \in \tilde{V}} p_v^{\alpha}(\tilde{v})\, q_{\mathcal{P},\alpha}^{1-\alpha}(\tilde{v}) \geq K \tag{4.13}$$

for all $v \in V$, with equality if $p(v) > 0$, where

$$q_{\mathcal{P},\alpha}(\tilde{v}) \overset{\text{def}}{=} \frac{1}{g_\alpha(\mathcal{P})} \Big(\sum_{v \in V} p(v)\, p_v^{\alpha}(\tilde{v}) \Big)^{\frac{1}{\alpha}} . \tag{4.14}$$

Then $K = [G(\alpha)]^{\alpha}$ and for any PD

$$\mathcal{Q} = \{ q(\tilde{v}),\ \tilde{v} \in \tilde{V} \} \neq \mathcal{Q}_{\mathcal{P},\alpha}$$

we have

$$\min_{v \in V} \sum_{\tilde{v} \in \tilde{V}} p_v^{\alpha}(\tilde{v})\, q^{1-\alpha}(\tilde{v}) < [G(\alpha)]^{\alpha} \tag{4.15}$$

or, equivalently,

$$\max_{v \in V} \frac{1}{\alpha - 1} \log_2 \sum_{\tilde{v} \in \tilde{V}} p_v^{\alpha}(\tilde{v})\, q^{1-\alpha}(\tilde{v}) > F(\alpha). \tag{4.16}$$

<u>Proof</u>. Since $\varphi(t) = t^{\alpha}$ is concave,

$$\sum_{v\in V} p(v)\, p_v^{\alpha}(\tilde{v}) \leq \Big(\sum_{v\in V} p(v)\, p_v(\tilde{v})\Big)^{\alpha} = q_{\mathcal{P}}^{\alpha}(v),$$

whence $0 < g_\alpha(\mathcal{P}) \leq 1$ immediately follows, implying $0 \leq f_\alpha(\mathcal{P}) < \infty$, as well.

Next we show that the function

$$\bar{g}(u) \overset{\text{def}}{=} \log_2 g_{\frac{1}{u}}(\mathcal{P}) =$$

(4.17) $$= \log_2 \sum_{\tilde{v}\in\tilde{V}} \Big(\sum_{v\in V} p(v)\, p_v^{\frac{1}{u}}(\tilde{v})\Big)^{u} \qquad (u > 0)$$

is convex and it is strictly convex if $f(\mathcal{P}) > 0$.

Suppose that $u = \beta u_1 + (1-\beta) u_2 \quad (0<\beta<1)$ and apply the well-known inequality (*)

(4.18) $$\sum_{i\in I} a_i^{\gamma}\, b_i^{1-\gamma} \leq \Big(\sum_{i\in I} a_i\Big)^{\gamma} \Big(\sum_{i\in I} b_i\Big)^{1-\gamma} \qquad (0<\gamma<1)$$

with the choice

$$I = V, \quad a_i = p(v)\, p_v^{\frac{1}{u_1}}(\tilde{v}), \quad b_i = p(v)\, p_v^{\frac{1}{u_2}}(\tilde{v}), \quad \gamma = \frac{\beta u_1}{u}.$$

We obtain, raising both sides to the power u,

$$\Big(\sum_{v\in V} p(v)\, p_v^{\frac{1}{u}}(\tilde{v})\Big)^{u} \leq$$

(4.19) $$\leq \Big(\sum_{v\in V} p(v)\, p_v^{\frac{1}{u_1}}(\tilde{v})\Big)^{\beta u_1} \Big(\sum_{v\in V} p(v)\, p_v^{\frac{1}{u_2}}(\tilde{v})\Big)^{(1-\beta)u_2}.$$

Now apply (4.18) with the choice $I = \tilde{V}$, $a_i = \Big(\sum_{v\in V} p(v)\, p_v^{\frac{1}{u_1}}(\tilde{v})\Big)^{u_1}$,

---

(*) A simple proof of (4.16) is : one may assume $\sum_{i\in I} a_i = \sum_{i\in I} b_i = 1$; then $\sum_{i\in I} a_i^{\gamma} b_i^{1-\gamma} = 1$ if $\gamma = 0$ or $\gamma = 1$, and since $a_i^{\gamma} b_i^{1-\gamma} = b_i\,(a_i / b_i)^{\gamma}$ is a convex function of γ, the inequality holds for $0 < \gamma < 1$. It follows, too, that the inequality (4.16) is strict unless a_i / b_i is constant.

$b_i = \left(\sum_{v\in V} p(v)\, p_v^{\frac{1}{u_2}}(\tilde{v})\right)^{u_2}$, $\gamma = \beta$; on account of (4.19), we obtain

$$\sum_{\tilde{v}\in\tilde{V}} \left(\sum_{v\in V} p(v)\, p_v^{\frac{1}{u}}(\tilde{v})\right)^{u} \leq$$
$$\leq \left(\sum_{\tilde{v}\in\tilde{V}} \left(\sum_{v\in V} p(v)\, p_v^{\frac{1}{u_1}}(\tilde{v})\right)^{u_1}\right)^{\beta} \left(\sum_{\tilde{v}\in\tilde{V}} \left(\sum_{v\in V} p(v)\, p_v^{\frac{1}{u_2}}(\tilde{v})\right)^{u_2}\right)^{1-\beta}. \tag{4.20}$$

Taking logarithms of both sides of (4.20),

$$\bar{g}(u) \leq \beta\, \bar{g}(u_1) + (1-\beta)\, \bar{g}(u_2) \tag{4.21}$$

i.e. the convexity of $\bar{g}(u)$ follows.

If $f(\mathcal{P}) > 0$, there exists $\tilde{v} \in \tilde{V}$ with $p_{v_1}(\tilde{v}) \neq p_{v_2}(\tilde{v})$ for some $v_1 \neq v_2$ with $p(v_1)\, p(v_2) > 0$. For such $\tilde{v} \in \tilde{V}$, the inequality in (4.19) is strict, cf. footnote [2)]; this implies strict inequality in (4.21), i.e., strict convexity of $\bar{g}(u)$. Since $\bar{g}(1) = 0$ and $\bar{g}(u) \leq 0$, if $u > 1$ (this follows from $g_\alpha(\mathcal{P}) \leq 1$), the convex function $\bar{g}(u)$ must be decreasing; in view of (4.17), this proves that $g_\alpha(\mathcal{P})$ increases as α increases. Moreover, the convexity of $\bar{g}(u)$ also implies that $\frac{\bar{g}(u) - \bar{g}(1)}{u-1} = \frac{\bar{g}(u)}{u-1}$ is an increasing function of u; substituting $u = \frac{1}{\alpha}$, and changing sign, in view of (4.9) and (4.17) we obtain that $f_\alpha(\mathcal{P})$ is an increasing function of α. If $f(\mathcal{P}) > 0$, the strict convexity of $\bar{g}(u)$ yields that $g_\alpha(\mathcal{P})$ and $f_\alpha(\mathcal{P})$ are strictly increasing.

The first equality of (4.11) is obvious from (4.9), being $g_1(\mathcal{P}) = 1$.

Carrying out differentiation (*)

$$\frac{\partial g_\alpha(\mathcal{P})}{\partial \alpha}\bigg|_{\alpha=1} = \frac{d}{d\alpha}\left[\sum_{\tilde{v}\in\tilde{V}} \exp\left\{\frac{1}{\alpha}\ln \sum_{v\in V} p(v)\, p_v^\alpha(\tilde{v})\right\}\right]_{\alpha=1} =$$

$$= \sum_{\tilde{v}\in\tilde{V}}\left(\sum_{v\in V} p(v)\, p_v(\tilde{v})\right)\left[-\ln \sum_{v\in V} p(v) p_v(\tilde{v}) + \frac{\sum_{v\in V} p(v) p_v(\tilde{v}) \ln p_v(\tilde{v})}{\sum_{v\in V} p(v) p_v(\tilde{v})}\right] =$$

$$= \sum_{\tilde{v}\in\tilde{V}} \sum_{v\in V} p(v)\, p_v(\tilde{v}) \ln \frac{p_v(\tilde{v})}{q_{\mathcal{P}}(\tilde{v})} . \qquad (4.22)$$

This proves (4.11).

Since $g_\alpha(\mathcal{P})$ and $f_\alpha(\mathcal{P})$ are increasing functions of α, so are $G(\alpha)$ and $F(\alpha)$, too. (4.12) is an immediate consequence of (4.10) and (4.11) being $f_\alpha(\mathcal{P})$ an increasing function of α.

If V and $\tilde{V}$ are finite sets, $f_\alpha(\mathcal{P})$ is a con-

---

(*) It is legitimate to change the order of summation and differentiation, even if V and $\tilde{V}$ are infinite sets. This is simple for the inner sums, since their terms are convex functions of α, thus, for the difference quotients, the monotone convergence theorem applies. As to the summation for $\tilde{v} \in \tilde{V}$ the terms are convex functions of $u = \frac{1}{\alpha}$ (this easily follows from (4.19)), hence the sum may be differentiated term-by-term with respect to $u = \frac{1}{\alpha}$ and then also with respect to α.

tinuous function of $\mathcal{P} = \{ p(v),\ v \in V \}$, hence its maximum is attained; let $\mathcal{P}$ be a maximizing PD. If $v_1 \in V$, $v_2 \in V$ are arbitrary with $p(v_1) > 0$, consider the function $h(t) = g_\alpha(\mathcal{P}_t)$, $0 \leq t < p(v_1)$, where $\mathcal{P}_t$ is obtained by substituting $p(v_1) - t$ and $p(v_2) + t$ for $p(v_1)$ and $p(v_2)$, respectively. Then, $h(t)$ is minimized for $t = 0$, i.e. $\left.\frac{dh}{dt}\right|_{t=0} \geq 0$. We obtain

$$0 \leq \left.\frac{dh}{dt}\right|_{t=0} = \sum_{\tilde{v}\in\tilde{V}} \frac{1}{\alpha}\Big(\sum_{v\in V} p(v)p_v^\alpha(\tilde{v})\Big)^{\frac{1-\alpha}{\alpha}}\Big(-p_{v_1}^\alpha(\tilde{v})+p_{v_2}^\alpha(\tilde{v})\Big), \tag{4.23}$$

which means

$$\sum_{\tilde{v}\in\tilde{V}} p_{v_1}^\alpha(\tilde{v})\, q_{\mathcal{P},\alpha}^{1-\alpha}(\tilde{v}) \leq \sum_{\tilde{v}\in\tilde{V}} p_{v_2}^\alpha(\tilde{v})\, q_{\mathcal{P},\alpha}^{1-\alpha}(\tilde{v}), \tag{4.24}$$

whenever $p(v_1) > 0$. This result proves the necessity of (4.13).

Suppose now, that $\mathcal{Q} = \{ q(\tilde{v}),\ \tilde{v} \in \tilde{V} \}$ is a PD on $\tilde{V}$ satisfying

$$\sum_{\tilde{v}\in\tilde{V}} p_v^\alpha(\tilde{v})\, q^{1-\alpha}(\tilde{v}) \geq [G(\alpha)]^\alpha \text{ for all } v \in V. \tag{4.25}$$

Multiplying by $p(v)$ and summing for all $v \in V$, (4.25) yields, on account of (4.14)

$$\sum_{\tilde{v}\in\tilde{V}} \left[g_\alpha(\mathcal{P})\, q_{\mathcal{P},\alpha}(\tilde{v})\right]^\alpha q^{1-\alpha}(\tilde{v}) \geq [G(\alpha)]^\alpha. \tag{4.26}$$

In force of the inequality (4.18), we have $\sum_{v\in V} q_{\mathcal{P},\alpha}^\alpha(\tilde{v})\, q^{1-\alpha}(\tilde{v}) \leq 1$, with strict inequality unless $\mathcal{Q} = \mathcal{Q}_{\mathcal{P},\alpha}$ (cf. footnote [2]). Thus, since $g_\alpha(\mathcal{P}) = G(\alpha)$, the assumption (4.25) leads to a contradiction if $\mathcal{Q} \neq \mathcal{Q}_{\mathcal{P},\alpha}$, proving (4.15) and the equivalent (4.16).

It follows, in particular, that if (4.13) is satisfied for any PD $\mathcal{P} = \{ p(v), v \in V \}$ and constant K, then, necessarily, $K \geq [G(\alpha)]^\alpha$. On the other hand, multiplying by $p(v)$ the equations $\sum_{\tilde{v} \in \tilde{V}} p_v^\alpha(\tilde{v})\, q_{\mathcal{P},\alpha}^{1-\alpha}(\tilde{v}) = K$ (valid, by assumption, if $p(v) > 0$) and summing for all $v \in V$, we obtain, on account of (4.14), $[g_\alpha(\mathcal{P})]^\alpha = K$. This proves both $K = [G(\alpha)]^\alpha$ and the sufficiency of the conditions (4.13).

<u>Remark.</u> The quantity

$$I_\alpha(\mathcal{P}_v \| \mathcal{Q}) \overset{\text{def}}{=} \frac{1}{\alpha - 1} \log_2 \sum_{\tilde{v} \in \tilde{V}} p_v^\alpha(\tilde{v})\, q^{1-\alpha}(\tilde{v})$$

may be interpreted as a generalized information-theoretic measure of the "distance" of the PD's $\mathcal{P}_v$ and $\mathcal{Q}$, called I - divergence of order α (of which the usual I - divergence is the limiting case $\alpha \to 1$). In view of lemma 4.2 the quantity $F(\alpha)$ has a similar geometric interpretation as the information capacity C^I, with the only difference that I - divergence of order α plays the role of usual I - divergence; thus, $F(\alpha)$ may be regarded as "information capacity of order α ". As to the numerical calculation of $F(\alpha)$, the same remarks apply as those concerning C^{I}.

Theorem 4.1. For a (simple) memoryless channel

$$\lambda(n,R) < 4\cdot 2^{-nE_r(R)} \qquad n = 1,2,\ldots \tag{4.27}$$

where

$$E_r(R) \overset{\text{def}}{=} \sup_{\frac{1}{2}\le\alpha\le 1} \frac{1-\alpha}{\alpha}\,(F(\alpha)-R) \tag{4.28}$$

is positive for $R < C$.

Here $F(\alpha)$ is meant with respect to the one-dimensional observation channel, i.e.

$$F(\alpha) \overset{\text{def}}{=} \sup_{\mathcal{P}} \frac{\alpha}{\alpha-1} \log_2 \sum_{\tilde{y}\in\tilde{Y}} \Big(\sum_{y\in Y} p(y)\,p_y^{\alpha}(\tilde{y})\Big)^{\frac{1}{\alpha}} \tag{4.29}$$

where $\mathcal{P} = \{p(y),\ y\in Y\}$ runs over the PD's on Y.

Proof. Apply the corollary of lemma 4.1 to the n - dimensional observation channel, setting $p(v) = \prod_{i=1}^{n} p(y_i)$ if $v = y_1\ldots y_n$. Since $p_v(\tilde{v}) = \prod_{i=1}^{n} p_{y_i}(\tilde{y}_i)$ if $v = y_1\ldots y_n$, $\tilde{v} = \tilde{y}_1\ldots\tilde{y}_n$, we have

$$\sum_{\tilde{v}\in\tilde{Y}^n}\Big(\sum_{v\in Y^n} p(v)\,p_v^{\alpha}(\tilde{v})\Big)^{\frac{1}{\alpha}} =$$

$$= \Big[\sum_{\tilde{y}\in\tilde{Y}} \sum_{y\in Y} p(y)\,p_y^{\alpha}(\tilde{y})\Big)^{\frac{1}{\alpha}}\Big]^{n}, \tag{4.30}$$

and from (4.3) follows (setting $N = 2^{nR}$)

$$(4.31) \qquad \lambda(n,R) < 2 \cdot 2^{\frac{1-\alpha}{\alpha}} \cdot 2^{n\left[\frac{1-\alpha}{\alpha} R + \log_2 \sum_{\tilde{y} \in \tilde{Y}} \left(\sum_{y \in Y} p(y)\, p_y^{\alpha}(\tilde{y})\right)^{\frac{1}{\alpha}}\right]}$$

This inequality holds for arbitrary PD $\mathcal{P} = \{p(y), y \in Y\}$ whenever $\frac{1}{2} \leq \alpha \leq 1$.

Thus, in view of $\frac{1-\alpha}{\alpha} \leq 1$ and of (4.29), also (4.27) is true. $E_r(R) > 0$ for $R < C$ follows from (4.12).

The function $E_r(R)$ is called the <u>random coding exponent</u> (since it has been derived by the method of random coding). In view of theorem 4.1, this provides a lower bound for the reliability function $E(R)$.

Using another approach referred to as the sphere-packing method, it may be shown that for memoryless channels with finite alphabets

$$(4.32) \qquad E_{sp}(R) \stackrel{\text{def}}{=} \sup_{0<\alpha\leq 1} \frac{1-\alpha}{\alpha} \left(F(\alpha) - R\right)$$

is an upper bound of the reliability function; $E_{sp}(R)$ is called the "sphere-packing exponent".

It is obvious from (4.28) and (4.32) that both $E_r(R)$ and $E_{sp}(R)$ are continuous, monotone decreasing and concave functions of R, positive for $0 < R < C$ and vanishing for $R \geq C$ (the latter statements follow from (4.12) and the fact that $F(\alpha)$ is increasing). $E_r(R)$ and $E_{sp}(R)$ coincide if $\frac{1-\alpha}{\alpha}(F(\alpha)-R)$ is maximized for $\alpha \geqq \frac{1}{2}$. Let us denote the infimum of such R's by R_{cr}.

Lemma 4.3. For $R \geqq R_{cr}$ we have

$$E_r(R) = E_{sp}(R). \tag{4.33}$$

Proof. We have to show that the parameter α for which $\frac{1-\alpha}{\alpha}(F(\alpha)-R)$ is maximized is an increasing function of R $(0 < R < C)$. For a fixed PD $\mathcal{P}$ we have

$$\frac{1-\alpha}{\alpha}(f_\alpha(\mathcal{P})-R) \stackrel{\text{def}}{=} -\log_2 g_\alpha(\mathcal{P}) - \frac{1-\alpha}{\alpha}R = -\log_2 g_{\frac{1}{u}}(\mathcal{P})-(u-1)R, \tag{4.34}$$

where $u = \frac{1}{\alpha}$. Here, according to the proof of lemma 4.2 (convexity of the function (4.17)) $-\log_2 g_{\frac{1}{u}}(\mathcal{P})$ is a concave function of u, hence the parameter $u = \frac{1}{\alpha}$ maximizing the right hand side of (4.33) decreases, i.e., α increases with increasing R. Since

$$\frac{1-\alpha}{\alpha}(F(\alpha)-R) \stackrel{\text{def}}{=} \sup_{\mathcal{P}} \frac{1-\alpha}{\alpha}(f_\alpha(\mathcal{P})-R), \tag{4.35}$$

it follows that the value α maximizing (4.35) also increases with R.

Thus, for memoryless (simple) channels

with finite alphabet, the exact form of the reliability function $E(R)$ is known for $R \geq R_{cr}$. For smaller rates the known upper and lower bounds $E_r(R)$ and $E_{sp}(R)$ do not coincide.

On the basis of lemma 4.1 it is possible to derive an upper bound $E_r(R)$ of the reliability function $E(R)$ also for channels with memory, in particular, for indecomposable finite state channels. It is believed that this bound is exact at least for rates close to capacity but for the time being no general proof of this exists.

5. Historical Notes and References.

The fundamental ideas that reliable transmission with a positive rate is possible over noisy channels and that the maximum of such rates may be calculated in terms of the measure of mutual information, are due to C. Shannon (1948). The first rigorous proof of the coding theorem for memoryless (simple) channels was given by A. Feinstein (1954); the strong converse is due to J. Wolfowitz (1957). The coding theorem for finite state channels is due to D. Blackwell, L. Breiman, A. Thomasian(1958). The significance of information stability was discovered by Dobrushin (1959), who has proved a very general version of the eoding theorem. The concept of compound channels was introduced by D. Blackwell, L. Breiman, A. Thomasian (1959) and J. Wolfowitz (1960). The reliability function for memoryless channels for $R \geq R_c$ has first been determined essentially by R. Fano (1961). The simple derivation of the bound $E_r(R)$ given in the text is due to R. Gallager (1965) while the sphere-packing bound $E_{sp}(R)$ was derived by C. Shannon, R. Gallager, E. Berlekamp (1967).

More complete references and much additional material may be found in the text-books listed below of which Gallager's excellent book is best suited for further study. For probability theory back-ground, we refer to Feller's book.

[1] Ash, R.B. Information Theory, Interscience Publishers, New York, 1965

[2] Fano, R.M. Transmission of Information, M.I.T. Press and Wiley, New York, 1961

[3] Feller, W. An Introduction to Probability Theory and its Applications, Vol. I, 2nd. Ed., Wiley, New York, 1957

[4] Gallager, R.G. Information Theory and Reliable Communication, Wiley, New York, 1968

[5] Jelinek, F. Probabilistic Information Theory, McGraw-Hill, New York, 1968

[6] Wolfowitz, J. Coding Theorems of Information Theory, 2nd. Ed., Springer, 1964.

Contents.

Printed in the United States
By Bookmasters